A CULTURAL HISTORY OF MATHEMATICS

VOLUME 1

A Cultural History of Mathematics
General Editors: David E. Rowe and Joseph W. Dauben

Volume 1
A Cultural History of Mathematics in Antiquity
Edited by Michael N. Fried

Volume 2
A Cultural History of Mathematics in the Medieval Age
Edited by Joseph W. Dauben, Clemency Montelle, and Kim Plofker

Volume 3
A Cultural History of Mathematics in the Early Modern Age
Edited by Jeanne Peiffer and Volker R. Remmert

Volume 4
A Cultural History of Mathematics in the Eighteenth Century
Edited by Maarten Bullynck

Volume 5
A Cultural History of Mathematics in the Nineteenth Century
Edited by Tom Archibald and David E. Rowe

Volume 6
A Cultural History of Mathematics in the Modern Age
Edited by Tom Archibald and David E. Rowe

A CULTURAL HISTORY OF MATHEMATICS

IN ANTIQUITY

VOLUME 1

Edited by Michael N. Fried

BLOOMSBURY ACADEMIC
LONDON • NEW YORK • OXFORD • NEW DELHI • SYDNEY

BLOOMSBURY ACADEMIC
Bloomsbury Publishing Plc
50 Bedford Square, London, WC1B 3DP, UK
1385 Broadway, New York, NY 10018, USA
29 Earlsfort Terrace, Dublin 2, Ireland

BLOOMSBURY, BLOOMSBURY ACADEMIC and the Diana logo are trademarks of Bloomsbury Publishing Plc

First published in Great Britain 2024

Series design by Raven Design
Cover image: Attic vase of Exekias depicting Achilles and Ajax playing dice, 6th century BC.
© DEA Picture Library / De Agostini via Getty Images

A catalogue record for this book is available from the British Library.

A catalog record for this book is available from the Library of Congress.

ISBN: HB: 978-13500-6293-1
ePUB: 978-13502-9475-2
ePDF: 978-13502-9474-5
Set: 978-13500-6312-9

Series: The Cultural Histories Series

Typeset by RefineCatch Limited, Bungay, Suffolk
Printed and bound in Great Britain

CONTENTS

ILLUSTRATIONS

FIGURES

Introduction

Chapter 1

Chapter 2

Chapter 3

Chapter 4

Chapter 5

Chapter 6

Chapter 7

Appendix

DIAGRAMS

Introduction

Chapter 3

Chapter 4

Chapter 5

TABLES

Chapter 3

Chapter 5

Appendix

TEXTS

Chapter 1

Chapter 2

CONTRIBUTORS

J. Lennart Berggren is Emeritus Professor of the History of Mathematics in the Department of Mathematics at Simon Fraser University, Canada. His special interests are the history of mathematical sciences in ancient Greece and medieval Islam, areas in which he has authored or co-authored numerous refereed papers. Among his eight books are *Ptolemy's Geography*, jointly with Alexander Jones (2000); and *Episodes in the Mathematics of Medieval Islam* (2016, 2nd ed.).

Eduardo A. Escobar is Assistant Professor of History of Science in the Department of Philosophy and Communication Studies at the University of Bologna, Italy. His research focuses on procedural texts and the transmission of technical knowledge in cuneiform scholarly cultures of ancient Iraq. He has published broadly on the history and historiography of scientific knowledge in the ancient world and is a core member of AlchemEast, a project funded by the European Research Council (ERC) on the early history of chemical recipes entitled.

Michael N. Fried is Chair of the Program for Science and Technology Education in the School of Education at Ben-Gurion University of the Negev, Israel. He is the author of *Apollonius of Perga's* Conica*: Text, Context, and Subtext* with Sabetai Unguru (2001); *Apollonius of Perga's* Conics, *Book IV: Translation, Introduction, and Diagrams* (2002); and *Edmund Halley's Reconstruction of the Lost Book of Apollonius' Conics: Translation from Latin, Introduction and Commentary* (2011).

Merav Haklai is Lecturer in Roman History at the Department of History, Ben-Gurion University of the Negev, Israel. Her main research focus lies with legal, economic, and monetary history of the Roman Empire. She has published on aspects of Roman monetary history, Roman economic behavior, and public trust, analyzing general patterns in Roman economic history and the interaction between legal traditions in operation within the Roman regime. Among her publications are, "Aurelian's Monetary Reform: Between Debasement and Public Trust" (2011); and "Credit and Financial Capital in Roman Egypt," in Paul Erdkamp, Koenraad Verboven, and Arjan Zuiderhoek eds., *Capital, Investment and Innovation in the Roman World*, Oxford Studies on the Roman Economy (2020).

Sir Geoffrey Lloyd is Emeritus Professor of Ancient Philosophy and Science at the University of Cambridge, UK, where he was Master of Darwin College from 1989 to 2000. His interests in ancient Greek and Chinese philosophy and science have increasingly taken him into comparative studies drawing on social anthropology, evolutionary psychology, ethology, and cognitive science, as most recently in his *Expanding Horizons in the History of Science* (2021).

Cécile Michel is Director of Research at the Centre national de la recherche scientifique (CNRS), Archéologies et sciences de l'Antiquité (ArScAn), Histoire et Archéologie de l'Orient Cunéiforme (HAROC), Nanterre, France, and Professor at Universität Hamburg, Centre for the Study of Manuscript Cultures (CSMC), Germany. An Assyriologist and member of the excavation team of Kültepe, Anatolia, her research focuses on the economy, society, women and gender, material culture, archival practices, literacy, education, and mathematics of the Ancient Near East. Among her recent publications is *Mathematics, Administrative and Economic Activities in Ancient Worlds*, co-edited with K. Chemla (2020).

Reviel Netz studied for his BA and MA in Tel Aviv University and received his PhD under the supervision of Geoffrey Lloyd at the University of Cambridge, UK. Since 1999, he has been teaching at the Department of Classics at Stanford University, CA, USA, where he is the Patrick Suppes Professor of Greek Mathematics and Astronomy. He is the author of many works in Greek mathematics and related fields, most recently *A New History of Greek Mathematics* (2022).

Christine Proust is Directrice de Recherche Emérite at the laboratory Sciences, Philosophie, Histoire (SPHERE), Centre national de la recherche scientifique and Université Paris Cité (CNRS), France. She is a historian of mathematics and her research focuses on the history and historiography of mathematics in the ancient Near East. She participated in the European Project, Mathematical Sciences in the Ancient World, 2011–2016, directed by Karine Chemla. Christine was awarded the Kenneth O. May Prize and Medal in the History of Mathematics. Among her publications are, *Tablettes mathématiques de Nippur* (2007) ; and *Scholars and Scholarship in Late Babylonian Uruk* (2019).

Francesca Rochberg is Catherine and William L. Magistretti Distinguished Professor Emerita in the Department of Middle Eastern Languages and Cultures at the University of California, Berkeley, USA. She specializes in the astronomical sciences in the ancient Middle East and Mediterranean. Most recently she has published *Before Nature: Cuneiform Knowledge and the History of Science* (2016); and co-edited with Alan C. Bowen *Hellenistic Astronomy: The Science in Its Contexts* (2020).

Liba Taub is Professor Emerita of the History and Philosophy of Science at the University of Cambridge, UK, where she is currently Director of Research in the Whipple Museum of the History of Science. Liba is also a Fellow Emerita of Newnham College, University of Cambridge. Having published extensively on ancient Greek and Roman science as well as the material culture of science, her most recent monograph is *Science Writing in Greco-Roman Antiquity* (2017). She is also the author of *Ancient Greek and Roman Science: A Very Short Introduction* (2023).

SERIES PREFACE

Culture, history, and mathematics are words that seldom appear together. Yet, mathematical activities have played a pervasive role in diverse cultures throughout human history, as described in the present six volumes. These illustrate how individuals and societies have engaged with various manifestations of mathematical culture from antiquity to the present. Based on recent and ongoing scholarship, they provide a global overview of a highly diverse range of developments, though certainly nothing like a comprehensive picture. Editors and authors had to make difficult choices about which topics and cultures to include, and in doing so, they have sought to strike a balance between characteristic features of a period and particular individuals and events that had a deep influence or longer lasting impact.

Each volume addresses seven general themes in as many chapters. The first three chapters —Everyday Numeracy, Practice and Profession, and Inventing Mathematics—highlight the roles of those who used, taught, or created mathematics as it evolved over time. These chapters explore mathematics per se, whereas the three that follow—Mathematics and Worldviews, Describing and Understanding the World, Mathematics and Technological Change—consider the larger impact of mathematics on other dimensions of human culture. The final chapter—Representing Mathematics—is devoted to a broader understanding of mathematical ideas, activities, or artifacts in different cultures throughout the world. The periods covered are:

Volume 1: A Cultural History of Mathematics in Antiquity
(3000 BCE–500 CE)
Volume 2: A Cultural History of Mathematics in the Medieval Age
(500–1400)

Volume 3: A Cultural History of Mathematics in the Early Modern Age (1400–1687)
Volume 4: A Cultural History of Mathematics in the Eighteenth Century (1687–1800)
Volume 5: A Cultural History of Mathematics in the Nineteenth Century (1800–1914)
Volume 6: A Cultural History of Mathematics in the Modern Age (1914–2020)

General Editors: David E. Rowe and Joseph W. Dauben

Introduction

MICHAEL N. FRIED

This volume concerns the cultural history of mathematics in the ancient world. Every word in that description has its subtleties and complexities. Culture, history, and mathematics, each by itself is far less clear than one might think, not to speak of the combinations, "cultural history," "history of mathematics," and even "ancient world." Part of the collective goal of the chapters in this volume is to explore the meaning of these words, most of all the meaning of mathematics itself in the context of the ancient world. Individually, though, the chief aim of the chapters will be to present concrete aspects of mathematics in ancient times in ways that will be new, informative, and eye-opening for the reader. More than in any individual chapter, however, the task of this introduction will be to take up directly the overall challenge of the volume and ask what it is we must think about when we consider a cultural history of mathematics in antiquity.

The questions raised by this volume, the first in the series, are shared to a great extent with Volumes 2–6, as one gathers from the identical chapter headings. This has the great benefit of inviting comparisons, but such comparisons raise further questions. For the subjects of culture, history, and mathematics do not merely differ in content from period to period, and from place to place, but also differ in meaning. And with those differences come differences in the meanings of our seemingly innocent chapter headings: what numeracy means in the context of ancient Iraq is something very different from what it means in nineteenth-century Britain; what a "profession" means, what technology is, will have vastly different answers if one is looking at Greece or Renaissance Italy or twentieth-century America. Understanding this fact alone goes far towards bringing out the difficulty and, accordingly, the great interest one will find in a cultural history of mathematics.

Putting this difficulty up front distinguishes this book from most other standard histories of mathematics. These standard works are written typically from a perspective that Ivor Grattan-Guinness colorfully called "heritage" as opposed to "history," meaning the viewpoint of the present looking back at the past (Grattan-Guinness 2004); they are written for the benefit of modern "inheritors" and most often only include mathematics that has led from the past to the present. Mistakes, dead ends, and false starts usually do not appear in such histories; motivations for mathematical work foreign to modern ways of thinking, as we shall see below when we discuss Proclus, are similarly ignored as unimportant. This series, in emphasizing a cultural approach to understanding mathematics and its history, by contrast, seeks to present the broader social context of mathematics, not as heritage but as history. Above all, nowhere do these volumes take for granted which kinds of activities should be called mathematical.

The problem of *mathematics* in the ancient world is a particularly difficult one since it is far from clear what mathematics *was* before there was a name for it, and even *after* the Greek name was adopted, as we shall see below. "Mathematics" obviously included less of what might be called mathematics today, but also included many things that are no longer considered to be "mathematics" at all (music, for example). This is all the more true for the ancient cultures of Mesopotamia, China, and India, which never used the Greek name for their own versions of "mathematics."

Mathematics seems to arise in all cultures from a basic observation of nature and various practical necessities—land measuring, surveying, taxation, building construction. But this raises other problems, if only because, unlike today, the separation between mathematics and its applications was not so sharply drawn in ancient times. For example, as Chapters 4 and 5 in this volume make clear, astronomy was *a kind* of mathematics rather than merely an application of mathematics. Likewise, mathematicians were called upon, as in ancient China, to determine auspicious or inauspicious days, and predicting comets or eclipses could foretell the fate of entire civilizations. Chapter 4, on mathematics and worldviews, points out that mathematics was more than a body of technical knowledge concerning measurement and calculation. As described there, "by providing access to the profoundest truths it contributed to happiness and fulfilment," so that "mathematics" also became a body of knowledge with moral and religious implications.

Adopting a cultural view of the history of mathematics means beginning with this indefiniteness as to what mathematics was taken to be. A cultural point of view also has implications for how the history of mathematics is written, its historiography. And historiography, far from being a mere technical matter, brings us back to epistemological questions about how one should think of mathematics: is it something eternal, essentially unchanging, perhaps only waiting to be discovered, or is mathematics a historical phenomenon, always changing, never quite defined, or rather always redefining itself? Is it one thing, or as varied almost as humanity itself? For now, we simply pose these questions for readers before proceeding further into the cultural history of ancient mathematics.

THE PROBLEM OF CULTURE AND A CULTURAL HISTORY OF MATHEMATICS

Of all the terms in our title, "culture" is, for sure, the most problematic. It is used everywhere and seems to mean everything so that one begins to wonder whether it actually tells us anything at all. In a well-known study of the idea of culture, the anthropologists A. L. Kroeber and Clyde Kluckhohn ([1952] 1963) presented and categorized some 164 definitions of "culture" and they had collected even more! A somewhat more updated work is Kuper's *Culture: The Anthropologists' Account* (Kuper 1999). In that book, Kuper comes to the extreme conclusion that we might do better "to avoid the hyper-referential word altogether, and to talk more precisely of knowledge, or belief, or art, or technology, or tradition, or even ideology" (p. x).

Besides a certain wooliness in thinking about "culture" magnified by the abundance of its meanings, the difficulty of "culture" may stem from several tensions in the concept itself. These include, first of all, the tension between what is natural and what is the result of human making or cultivation: with that, there is a further difficulty in culture as the ground of human activity and somehow also a human attainment, as when we speak about "high culture." Where "culture" is used in connection to mathematics, this first tension is particularly acute, not the least because, for many, mathematics is indeed more natural than cultural.

A different tension in the idea of culture has to do with whether culture begins with an individual or whether an individual is constituted by a culture. This, too, has some bearing on how we conceive our cultural history of mathematics, for example, whether this is a history of individual mathematicians or of intellectual currents that only swept individuals along with it. This, among other things, will be an issue in Chapter 2, "Practice and Profession" and in Chapter 3, "Inventing Mathematics," which will take up in terms of the ancient Near East and Greece the identity of a mathematician as an author or as a specialist with well-defined role in a social hierarchy.

We cannot go into any great detail about these particular matters in this introduction.[1] However, we do need to say more about an aspect of culture that is arguably most characteristic of it and one that leads to its own difficulties, namely, its diversity. For us, the mere mention of the word "culture" brings to mind *other* cultures, even if we speak about some culture that we deem our own.

A cultural history, therefore, refers to history from the perspectives of a variety of cultures—and, in what follows, the Greeks, the Egyptians, the Babylonians, the Chinese, the Indians, etc., are referred to as different "cultures." This immediately raises the question, How many cultures are there? How is one to distinguish one culture from another?

One might try to avoid the issue of diversity by restricting oneself to a single culture, Greek culture or Chinese culture, born in some particular place. This is not an unreasonable approach and is usually unavoidable. On the other hand, is Greece in the sixth century BCE, the Greece of Pythagoras, the same culture as Greece in the first century CE? Is it even possible to speak of "Greece" or the "Near East" as a single place or region having a distinctive culture? Where exactly was "Greece"? And where was the "Near East" or probably more precisely the ancient "Middle East"? In the case of China, this becomes even more problematic, for even after it was unified under the first emperor of China, Qin Shihuang, in 221 BCE, the previous "warring states" remained specific subcultures and the empire covered a vast territory including many regional nationalities with their own languages and customs. Is China one culture or many?

Even if a region is regarded as the seat of a specific culture, there is the question of subcultures. In Mesopotamia, for example, there was a class of people, the scribes, a class at the heart of mathematics in the Near East, whose world was different in crucial ways from that of other ordinary people in Mesopotamia. For this reason, one speaks of a scribal culture. This will be discussed in great detail in all the chapters focusing on the Near East, Chapters 2, 3, 5, and 6. But we may already ask, in what sense is this scribal culture a *genuine* culture, where does it begin and end with respect to a more general Near Eastern culture? And more generally throughout the volumes is this series, do mathematicians collectively form a subculture of whatever other culture it may be to which they belong?

The question of whether there is a single culture that has developed or many cultures with their own developments is one that readers will have to consider. In treating the diversity of culture, one might try to examine as many cultures as possible. But this can only lead to superficiality. In this volume, we chose to focus on two very long-lived and immensely influential sets of cultures, those of Classical world and those of Mesopotamia. These are different enough and developed enough to get a feeling for how diverse cultures can be. Of course, we are only interested in one aspect of these two sets of cultures, their mathematical activity. And the chapters of this volume—not to speak of the other volumes of the series!—will make it apparent that there can be real cultural differences in mathematical thought, differences that cannot simply be ascribed to periods of development as if the development of mathematics were a straightforward linear process in which before and after are clearly defined.

There are, for example, cuneiform astronomical texts that were written *at the same time* as Greek Hellenistic texts, each still maintaining their distinct ways of thinking. The one cannot be viewed necessarily as an earlier stage of the other.

To be sure, there was interchange between peoples about ideas, about mathematics and science, and influence from one period of time to another. This was certainly so regarding the two sets of cultures we are focusing upon, Greece and Rome and Mesopotamia and the ancient Near East. It is well known, for example, that the great astronomer Hipparchus (*c*. 190–120 BCE), whose work formed the foundation for Ptolemy's Almagest, used Babylonian observations and was, generally, profoundly influenced by the Babylonian astronomical work. Not only did one influence the other, both the Classical world and Mesopotamia have left deep footprints in cultures that came after them, including our own. As for the former, one only has to think of the axiomatic-deductive frameworks that characterize mathematics to this day, as almost symbolized by Euclid's *Elements* (see below). As for Mesopotamia, one only has to look at the clock, for our measurement of time and angles by way of divisions of 60 is an inheritance from the cuneiform cultures of Mesopotamia (see Chapter 3 for an account of the sexagesimal counting system, the base 60 number system).

But these similarities in the mathematical activity of different cultures and their influence on our own notwithstanding, we need to avoid a picture of the history of mathematics in terms of an overarching research program of which all these peoples were participants, as tempting as that may be. The diversity of cultures is always a diversity of thought. That does not mean, at the same time, that when one engages with a cultural history of mathematics one has to slide down the slippery slope towards a simplistic kind of cultural relativism in which one is forbidden to speak about truer, deeper, or better. Cultural historians do not reject 2+2=4 as a fact! Their basic presupposition is that thought at one time or place *about* that equation—even whether one writes such a thing as an equation—may be quite different from how *we* think about it. The number two, just to follow the example, may be the number two everywhere, but what a number is, what its meaning is, what role it has in mathematical thought, have all varied and changed.

So, while there may well be something like universal mathematical truths, the *meanings* attached to them differs depending on the human context of their development. In Chapters 5 and 6, for example, the reader will learn that scribes used certain numbers as "coefficients" in various calculations. Although historians have chosen to use the common mathematical term "coefficient" for these numbers, there is nothing common about them. Initially, one is tempted to compare these "coefficients" to physical constants or units, yet one quickly discovers how far that misses the mark. This is a real instance of where one might seriously misinterpret Mesopotamian mathematics by searching for an all too easy bridge to something familiar to present day mathematics. In struggling to understand what the ancient scribes meant by such numbers, it is necessary to appreciate that theirs was a truly different culture whose ideas and practices were not merely translations of our own.

These general issues we have described to this point—culture as ground or attainment, culture and individuals, diversity and relativism—even if they are not discussed explicitly, will always be in the background of the chapters below as will be many questions derived from them. How was mathematics passed on and what roles did it play in everyday life? How was it viewed by people, what kind of activity was it for them? The chapters on everyday numeracy, the profession, representing mathematics, and even the chapter on technology all touch on these questions. How does the human production of mathematics relate to nature or to the world as a whole? Such questions will be at the heart of the chapters on worldviews and mathematics and the world. How and why are mathematical ideas created sometimes in the same way, sometimes in different ways, in different human societies? Does this happen collectively or is it inspired by individuals? This last, as we already mentioned, is very much a part of the chapters on inventing mathematics and on practice and the profession.

Questions such as these lie at the heart of a cultural history of mathematics. This does not mean that this is history of mathematics without mathematics. Of course not. This series, however, is focused on understanding past mathematical cultures in their own distinctness: it is not driven by a goal to identify what did or did not contribute to contemporary, modern mathematics.

It is not an utterly illegitimate enterprise to look back at the past and try to understand how mathematical activity at another time and place reflects what is familiar today. In doing so, one may gain some useful insights about contemporary modern methods and ideas. To look at Archimedes' *Quadrature of the Parabola* in terms of today's concepts of integration, for example, which are basic to modern mathematics, may shed light on the kinds of conceptual difficulties and achievements that led to the calculus as we know it. One cannot deny this is *a* way of relating to the past, but again, it is one aligned with "heritage," *not* history, to use Grattan-Guinness' terms. And as to that, Grattan-Guinness adds in cautionary tone, "Heritage addresses the question 'how did we get here?,' and often the answer reads like 'the royal road to me'" (2004: 165).

What a cultural history tries to do instead is to understand what a certain kind of activity that we call mathematics meant for its practitioners and for the societies in which they lived. One would like to know what Archimedes' work meant to him and the others around him, naturally those who shared his interests, but also those who did not. In the eyes of the comic playwright Aristophanes, for one, mathematicians and other philosophers (for mathematicians were regarded as philosophers, in a way) were good-for-nothings to be laughed at, as we shall see in Chapter 7, "Representing Mathematics." This, too, is part of the cultural history of mathematics. One should not get the impression that the division is always clear between a cultural history of mathematics and a history which is heritage, or what may be called a mathematical history of mathematics; it is sometimes very unclear. There is, in fact, a spectrum of relationships to the mathematics of the past, each in their own way valid (see Fried 2018), and readers should keep in mind these other ways of possibly relating to ancient mathematics. But now, having looked briefly at culture and history, we need to look more closely at the difficulty of the term "mathematics" in ancient times.

MATHEMATICS

In his book, *Mathematics as a Cultural System*, Raymond Wilder, himself a first-rate mathematician, could write with some authority that: "Mathematicians no more agree on the nature of mathematics than do anthropologists in regard to culture" (1981: 11). The problem is all the more challenging when it comes to *ancient* mathematics, where one must often ask whether something really is mathematics at all.

The *word* mathematics is Greek. Other cultures have comparable terms. Typically, they are related to number, calculation, or reckoning. For example, the ancient Chinese term for mathematics, *suan shu*, means literally "the art of computation." And traditionally, Chinese mathematics arose from problems of practical consideration and was concerned with providing methods or algorithmic procedures for solving problems in set, formulaic ways. In India, the ancient Sanskrit term *gaṇita*, derived from *gaṇa* "multitude," likewise evokes fundamental practices of enumerating and reckoning.

The Greek term is somewhat broader in meaning and has an origin with no immediate reference to calculation or number. *Mathēmatika* comes from the Greek verb *manthanō* from which also the words *mathēma* and *mathēsis* are derived. All of these words have to do with learning: *manthanō*, "I learn," *mathēsis*, "an act of learning," and, *mathēma*, "a thing learned." So, etymologically speaking, "mathematics" has little to do directly with what we generally think of as mathematics. It was apparently the Pythagoreans (see Chapters 4 and 7) who gave the word its association familiar today as identifying those things having to do with number and shape with what is eminently learnable. By Plato's time both meanings still existed. Thus, in Plato's *Meno*, Meno uses a word derived from *manthanō* at the start of the dialogue when he asks whether virtue is something learned. But when Socrates wants to demonstrate to Meno what *learning* is, he does so by way of a *mathematical* lesson to an uneducated boy. Quite a few readers might balk, no doubt, at the claim that mathematical things are precisely those which can be learned! That notwithstanding, it is with this essentially philosophical claim connected to the learnability of numbers, magnitudes, and shape that mathematics becomes "mathematics."

But before the word "mathematics," before Plato's philosophical position was ever set out, should those activities that today are recognized as mathematical be called "mathematics"? Should a Greek name be given to activities engaged in by people who were not Greek, who did not necessarily share Plato's perspective or that of the Pythagoreans, who may not have grouped these activities under one concept, or who may have grouped them together with other "non-mathematical" activities? Should a common term be sought altogether? This is a difficult question. It is a question that needs to be asked, even if not answered, in a cultural history such as this one.

Consider the difficulty in terms of what is thought to be one of the earliest artifacts connected to mathematics, the Ishango bone. It is an example of a "tally," a stick or stone or, in this case, a bone, with marks or notches meant to record anything that might be counted (see Everett 2017). Is this where one should look for the origins of recorded numbers? The very oldest of such tallies is the Lebombo bone, produced some 43,000 years ago (Beaumont and Bednarik 2013: 37). It contains 29 distinct notches, perhaps related to the length of the lunar month. The Ishango bone is more recent, about 20,000 years old, and is a much more sophisticated tally than the older Lebombo bone. It was found in 1950 by a Belgian geologist-archaeologist Jean de Heinzelin at a site known as Ishango along the banks of Lake Edwards on the Uganda–Congo border. The bone is about 10 cm long and has tally marks notched along three sides (see Figure 0.1), as sketched in Diagram 0.1.

What strikes one immediately is the *grouping* of notches, so while it may be a tally, it is a complex one or, perhaps, a collection of tallies. Several patterns can be discerned in these groups of notches. De Heinzelin was struck by the presence of the primes 11, 13, 17, and 19 and thus attributed to the ancient people of Ishango a knowledge of prime numbers. More recently, while rejecting the Ishango people had knowledge of prime numbers, Pletser (2012) argued that the bone showed evidence of calculations with base 12 and sub-bases 3 and 4. While de Heinzelin and Pletser's give purely numerical contexts for the bone, other interpretations focus *on what* was being counted. One of the possibilities was that the Ishango bone was a kind of lunar calendar (Marshack 1991). This led Claudia Zaslavsky (1992) to suggest that it was, in fact, produced by a woman to follow menstrual cycles, and to conclude boldly that, "women were undoubtedly the first mathematicians!"

Such statements are as problematic as they are bold. The first thing, naturally, is that there are simply not enough numbers to establish a pattern. The British mathematician, Richard Guy, once wrote a paper called, tongue-in-cheek, the "Strong Law of Small Numbers," in which he provided "35 examples of patterns that seem to appear when we look at several small values of n" (Guy 1988: 697).

FIGURE 0.1. Ishango bone. Institut Royal des Sciences naturelles de Belgique, Bruxelles. Photo by Wikimedia Commons.

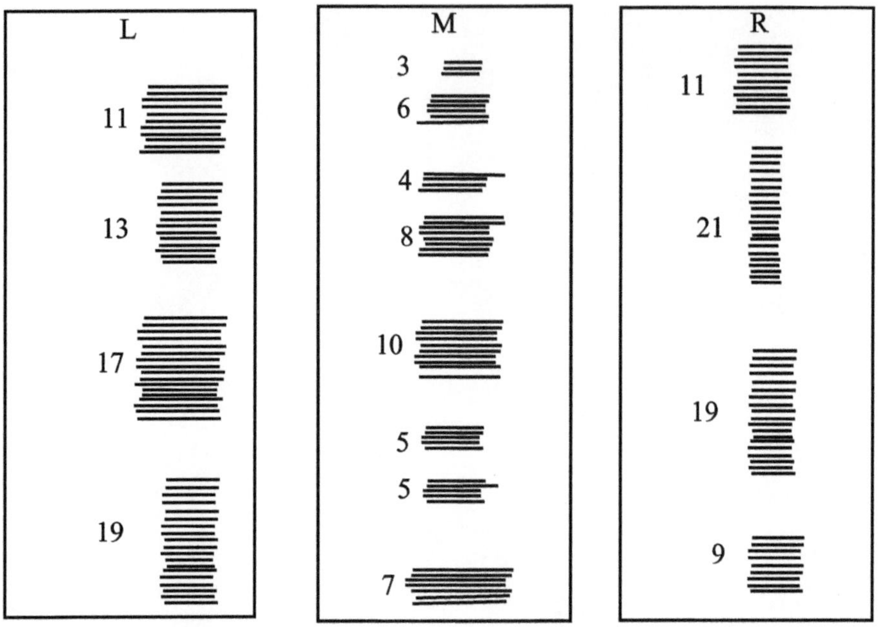

DIAGRAM 0.1. The notches along the sides of the Ishango bone.

It is quite easy to be seduced by such patterns. But assuming there is a pattern in an artifact like the Ishango bone, can we assume that the creator of the artifact, he or she, intended the same pattern that *we* discern?

What alone seems plain is that the lines are tallies, they are counting something, and that these counts were recorded. Still, granting that such tallies are ancient examples of recorded signs with numerical meaning, is it right to call this "mathematics"? Is it right to call whoever notched those bones a "mathematician"? Exactly what kind of activity does the Ishango bone represent?

These questions are no less relevant for ancient China where actual numbers, not just notches, appear on ox scapulas and turtle plastrons. Some of these date as early as 1200 BCE and were used for divinatory purposes during the late Shang dynasty, the first recorded Chinese dynasty. Chinese numerals are derived from Shang dynasty symbols going back to the fourteenth century BCE for numbers 1 through 9, and separate symbols for powers of 10, 100, 1000, making it possible to write even very large numbers on the oracle bones (see Figures 0.2a–b).

The situation is similar in the case of early Egyptian hieroglyphic number signs. Like the Chinese system, the Egyptian hieroglyphics used separate signs for 1, 10, 100, 1,000, 10,000, 100,000, and 1,000,000 as can be seen in Diagram 0.2, adapted from a *c.* 2300 BCE Old Kingdom funerary inscription

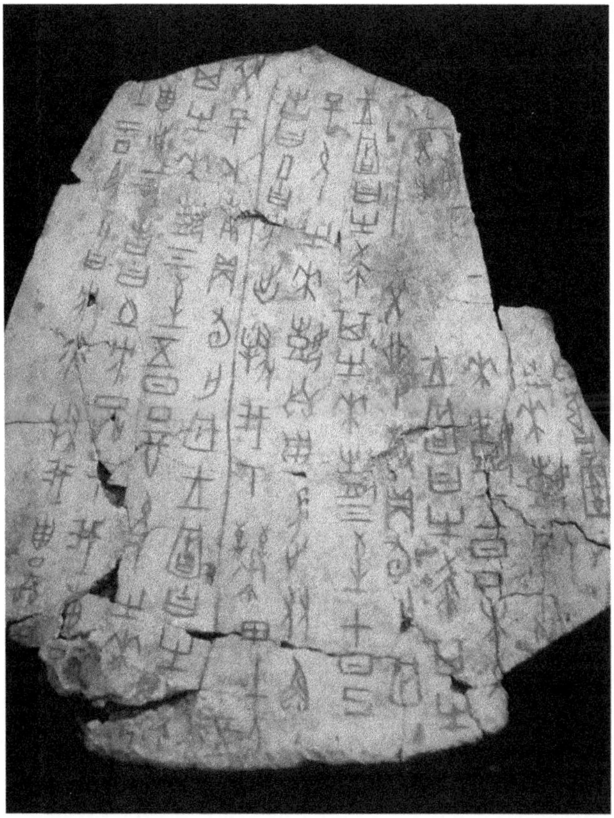

FIGURE 0.2a. Shang dynasty oracle bone showing oracle script numerals. Photo by Zens Photo via Getty Images.

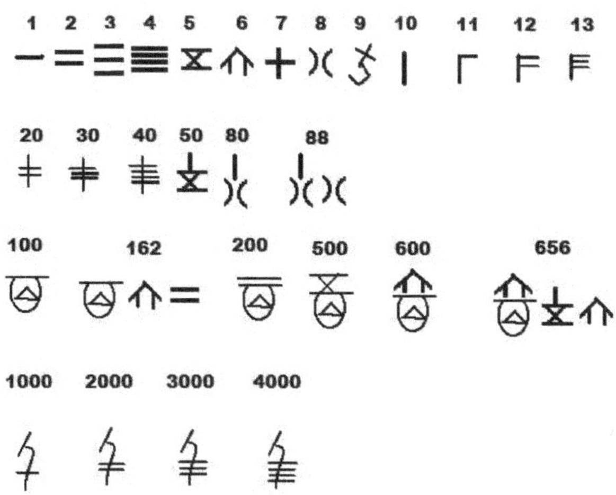

FIGURE 0.2b. The Shang dynasty numbering system. Photo by Wikimedia Commons.

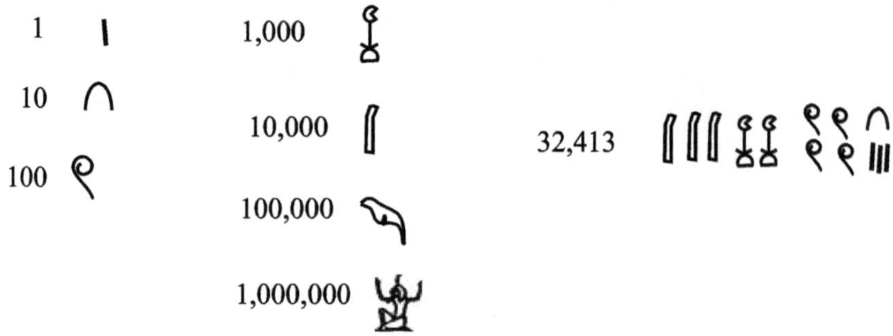

DIAGRAM 0.2. The Egyptian Numbering System (Old Kingdom).

(see Ifrah 1998: 166). As can be seen, there were no special signs for 5,6,7.8, or 9, nor for 500, 600, and so on. One repeated a sign as many times as needed up to nine times. The position of the signs, though following some loose conventions, had no essential role, as we shall see was the case in cuneiform numbering (see Chapters 2 and 3). Although there was clearly a kind of decimal number system here as there was with the Chinese, the repetitions make the act of recording numbers a little like the tallying represented by the primitive Ishango bone. Indeed, while the choice of signs for 10, 100, etc. are not completely clear, Annette Imhausen remarks (2016: 18) that the unit sign possibly originated in tallying sticks like the Ishango bone.

Then we have cases where no number signs are written explicitly yet numbers are present implicitly, even sophisticated number systems and calculations. For example, we have from as early as the third millennium BCE urban Indus Valley societies using elaborate scales of weight units (see Figure 0.3) and other metrological systems (see Danino 2014).

Even so, does the presence of such numbers, even hints of a number system, constitute "mathematics"? Should all activities or artifacts that involve numbers or counting be regarded as "mathematics"? Should we call all activities involving the production of geometrical shapes, for example, in the designs on pottery, mathematics? Does the Inca method of record keeping by tying knots in ropes constitute mathematics? Is an Egyptian scribe writing down the number sacks of grain doing mathematics? These are genuine, not rhetorical, questions. They are relevant to a good part of ancient mathematics but particularly relevant in connection to prehistoric artifacts when we are as close as we can get to the beginnings of mathematics (if we can call it that).

Part of the problem is that the world of human activity, even now, is not always easily articulated; in the ancient world especially it is not always easy to say this is "mathematics," this is "astronomy," this is "religion" (see Chapter 4). Nevertheless, we can say this much. There are certain cases in which numbers

FIGURE 0.3. Weights from ancient Harrapan period (2600-1900 BCE), which enabled trade and economy to flourish in the Indus Valley. These weights used a binary system that multiplied the smallest weight (0.85 g) by two, and then doubled each of the successive weights. This was exactly like the ancient Egyptian method of multiplication by doubling a given number. Weights such as these have been found in ancient cities like Mohenjo Daro, Chanhu Daro, and Harappa, often near city gates where such taxes would have been collected. National Museum, New Delhi Creative Commons. Photo by Wikimedia Commons.

and shapes, as well as the relationships among them, become objects of reflection: it is here one can begin to speak confidently about an activity being "mathematical." Obviously, one cannot know whether an ancient artifact, such as the Ishango bone, truly represents any such reflection, something that should urge caution in making pronouncements about the "first mathematicians." But there are instances in which such interest and focus is impossible to deny. For example, in Chapter 5, a Babylonian scribe enters this cultural history of mathematics because he was occupied in writing down the ungodly number 9^{46}! What for?! The number could hardly have had any practical use. Writing it down would surely have been a tremendous challenge and proof of numerical prowess: this can truly be called a "mathematical activity."

But even in the case of writing down a very large number, one must be cautious in guessing the *reason* behind the scribe's activity. Even where one finds suggestion of mathematical reflections, one must be careful not to assume what form these reflections took and what their real focus was. While such caution is especially fitting when looking at the very ancient period, it is equally fitting for later times as well. And the Greek world, which had the term "mathematics" at its disposal, is no exception. Consider Proclus' (5th century CE) view on what doing mathematics involved. In the initial prologue to his commentary on the first book of Euclid's *Elements*, he wrote the following referring to the word "mathematics" and its meaning:

> This, then, is what learning (*mathēsis*) is, recollection of eternal ideas in the soul; and this is why the study that especially brings us the recollection of these ideas is called the science concerned with learning (*mathēmatikē*). Its name thus makes it clear what sort of function this science performs. It arouses our innate knowledge, awakens our intellect, purges our understanding, brings to light the concepts that belong essentially to us, takes away the forgetfulness and ignorance that we have from birth, sets us free from the bonds of unreason . . . (Morrow [1970] 1992: 38)[2]

For Proclus, who was a Platonist, mathematics was a science that turned souls towards eternal things and ultimately towards the Good, or as he put it a few lines later in the same passage, "to the blessed life." Of course, not *all* of his commentary on Euclid concerns contemplation of the Good. The main part of the work discusses the propositions of Book I of the *Elements* and other aspects of mathematics more familiar to the modern reader. But here, too, there are surprises. For example, consider the parts of mathematics.

Proclus begins his account of the parts of mathematics with the Pythagoreans, who, as noted earlier, used the word "mathematics" specifically in connection with numbers, magnitudes, and shapes. Like the word "mathematics" itself, their classification scheme at first sounds familiar, or at least not quite so foreign to our modern ears as do references to a "blessed life." The scheme is based on the kinds of "mathematical things," as the ancient Pythagoreans saw them. To start, they divided things initially according to two basic categories, *magnitude* (*pēlikon*) and *multitude* (*poson*). The first included all of those things answering the question "How much?" things such as lengths, areas, volumes, and weights; the second included all of those things answering the question "How many?" The study of these mathematical things was further divided as follows. The study of magnitudes at rest was *geometry*, while that of magnitudes in motion, *astronomy*. When the study of multitudes, which was the study of numbers (*arithmoi*), considered numbers in and of themselves, this was *arithmetic* (*we* would say *higher arithmetic* or *number theory*); when it looked at numbers in relation to other numbers, the subject was *music*! These divisions are summarized below:

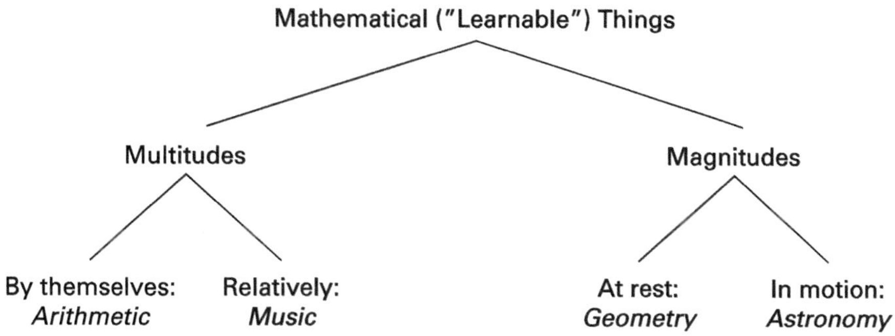

What is important to stress here is that besides the surprise of music being one the fundamental mathematical sciences, neither it nor astronomy were thought of as *applications* of mathematics. Like geometry and arithmetic, they are mathematical because they reflect modes of being, as it were, with respect to magnitudes and multitudes.

This fourfold division, we should mention, was important not only in the history of mathematics but also in the history of education. For education, from late antiquity, through the Middle Ages, and until roughly the Renaissance, was dominated by the idea of the Liberal Arts. These are the same arts referred to when one receives a degree of Bachelor of Arts or Master of Arts. The Liberal Arts were divided into two groups. The first, the *trivium*, comprised the three arts of expression, which were grammar, rhetoric, and logic. The second, the *quadrivium*, comprised the mathematical arts, which were precisely the Pythagorean sciences of arithmetic, music, geometry, and astronomy.[3] (This liberal arts tradition with its Pythagorean roots will be discussed briefly in Volume 2, Chapter 2, and again in Volume 3, Chapter 7.)

The Pythagorean scheme for mathematics, however, was not the only one in ancient times. Having discussed that scheme, Proclus mentions a different way of dividing "mathematical things"—one described in a work by the Greek astronomer and mathematician Geminus (1st century BCE). The division here is made according to whether one is concerned with objects of pure thought, intelligibles (*noēta*), or objects of the senses (*aisthēta*). The study of arithmetic and geometry belonged to the first group. In the second group, Geminus included six sciences: mechanics, astronomy, optics, geodesy, music, and calculation (*logistikē*). Again, although we recognize these subjects as somehow mathematical (with the exception perhaps of music), they were not regarded as *applications* of mathematics, they *were* mathematics. They were distinguished only by the nature of their mathematical objects.

The line between mathematics and applications of mathematics becomes much firmer in later times. Thus, in Volume 4 of this series, one will read about how the use of mathematics becomes wrapped up with technology and industrialization in the eighteenth century (see Volume 4, Chapter 5) and more dramatically, perhaps, in the nineteenth century with the advent of the Polytechnic movement for which mathematics and its application to engineering becomes a major theme (see Volume 5, Chapter 5).

This distinction, or rather lack of distinction in ancient times, between mathematics and its applications is crucial in order to understand the ways in which mathematics in antiquity becomes connected to "the world" as discussed in Chapters 4 and 5, and to some degree in Chapter 6 on technology as well. To be sure, one can read Ptolemy's book on astronomy, the *Mathematikē Syntaxis*, better known as the *Almagest*, as a book about astronomy that uses mathematics, especially if one thinks of it as a book comparable to a modern book on positional astronomy. But, if one is interested in a cultural, historical point of view, one would want to know how Ptolemy himself viewed his work. For Ptolemy, this was mathematics itself. His own description from the preface to the *Almagest* is reminiscent of the Pythagorean description of astronomy, but is striking in that Ptolemy mentions no other mathematics and only distinguishes what he is doing from other kinds of theoretical *philosophy*:[4]

> That division [of theoretical philosophy] which determines the nature involved in forms and motions from place to place, and which serves to investigate shape, number, size, and place, time and suchlike, one may define as "mathematics." (Trans. from Toomer 1984: 36)

Ptolemy then goes on to say that mathematics falls between physics and theology in concerning both sensible things (as Geminus had also taken into account) and eternal, divine things. He even moves into what might be considered non-mathematical realms of a spiritual and even ethical kind. Towards the end of his preface, Ptolemy writes:

> With regard to virtuous conduct in practical actions and character, this science [i.e., mathematics generally and astronomy in particular], above all things, could make men see clearly; from the constancy, order, symmetry and calm which are associated with the divine, it makes its followers lovers of this divine beauty, accustoming them and reforming their natures, as it were, to a similar spiritual state. (Toomer 1984: 36–7)

MATHEMATICS AND THE HISTORIOGRAPHY OF MATHEMATICS

One might suggest that ignoring such remarks about "virtuous conduct" and "divine beauty" and taking advantage of the power of modern mathematics to abstract and look beyond these strange idiosyncratic formulations, one might reach the "true" core of Ptolemy's or Euclid's thought. But, recalling the distinction drawn earlier between heritage vs. history, one can ask whether this would really be history, even though it has to do with the past. For, as we have already discussed in the first section, it is a central aim of a cultural history such as this one to explore the very idiosyncrasies that that generalizing power of modern mathematics sweeps away. Indeed, if we take seriously what Ptolemy *actually* wrote, we have to accept the possibility that "virtuous conduct in practical actions and character" *was* at the core of his thought as *he* saw it.

The question whether modern mathematics can be used as a legitimate tool to understand ancient mathematics, or the mathematics of any earlier period, is a question to be taken seriously. The underlying question is what allows us to arrive at true *historical* understanding of the mathematics of the past. This is a matter of historiography, the general inquiry into the methods and approaches of historical research and, by implication, the nature of historical knowledge. In all areas of history, historiographical questions are of great interest to historians; but they are nowhere more controversial and vexed than in the history of mathematics, especially of ancient mathematics.

A point of some general agreement in historiography is that anachronism—the bringing of modern conceptions and concerns to bear on considerations of the past—always risks distorting our view of the past. That is the mild version; the more extreme one is that anachronism is fatal to historical understanding, plain and simple! When it comes to mathematics, it is undeniable that the proof of a geometrical theorem in Euclid, for the most part, remains valid today, so is it necessary or helpful to put aside today's understanding of geometry, and all the richness of non-Euclidean geometries, for example, when reading Euclid? In other words, when it comes to mathematics, is it possible that anachronism might even be in some ways helpful, if not enlightening?[5] The problem again, however, is whether it would be enlightening in terms of the history of mathematics, and in particular, of a *cultural* history of mathematics.

It is not by accident that Euclid (*fl. c.* 300 BCE) is mentioned in this connection. One of the most well-known debates about the historiography of mathematics—and it was an acrimonious one—concerned an interpretation of Book II of Euclid's *Elements*. At the heart of the debate was precisely the question of anachronism. Without going into details, for much has been written about it (e.g., Berggren 1984; Fauvel and Gray 1987; Schneider 2016), the controversy began with a paper by a historian of ancient mathematics, Sabetai Unguru, which had the provocative title: "The Need to Rewrite the History of Greek Mathematics" (1975). The title made it clear that the problem in Unguru's view went beyond the particular text which was the article's ostensible concern, Book II of Euclid's *Elements*. Heated responses came from eminent mathematicians such as Bartel Leendert van der Waerden, Hans Freudenthal, and André Weil. It was very much a battle between a mathematical versus a historical point of view, whether history of mathematics was history or mathematics (see Unguru 1979: 563). Although the central points of Unguru's historiographical position were eventually accepted, there continue to be voices adding nuance to the discussion, and, on occasion, a revival of polemics (e.g., Blåsjö, 2016).

Whatever else may be said about this debate, it demonstrates both the tremendous attraction of an anachronistic reading of a mathematical text, owing to the power of modern mathematics to find general abstract patterns, but also its dangers and distortions. Instead of discussing in a theoretical manner the sometimes overt and sometimes subtle ways those attractions and dangers present themselves, it would be best to see how these issues can arise even in a mere overview of a book like the *Elements*, a book that, for many, is the very paradigm of a mathematical text, and a book that has had a persistent presence in the history of mathematics, even in China, a book read, copied, revised, and debated until the twentieth century. Needless to say, readers will feel its presence throughout the volumes of this series.

EUCLID'S *ELEMENTS*

Besides the *Elements*, Euclid also wrote works on spherical astronomy, optics, and music, a work on conics, on "surface loci," on dividing figures, on [geometric] fallacies, among others, and a curious work called the *Data, Dedomena* in Greek, which covers material also found in the *Elements*, but from a very different standpoint. As is the case with much of Greek mathematics, many of these works have been lost. What Euclid wrote about conics, on surface loci, on fallacies, and others have never been seen by modern eyes. These are known only through accounts by later figures such as Proclus and the fourth-century CE mathematician, Pappus of Alexandria (see Chapter 7 for more about Pappus), whose *Collection* was one of the great syntheses of Greek mathematics in late antiquity. Indeed, it must be said that before Euclid there are no real extant Greek mathematical texts, and even what survives of Euclid is through diligent (and some not so diligent) copies made of his work. Although fragments, some even with diagrams, have come down to us from Roman times, such as the Egyptian papyrus in Figure 0.4, *c*. 75–125 CE, the earliest surviving copy of the *Elements* dates from the ninth century CE, more than a millennium after he wrote it.[6]

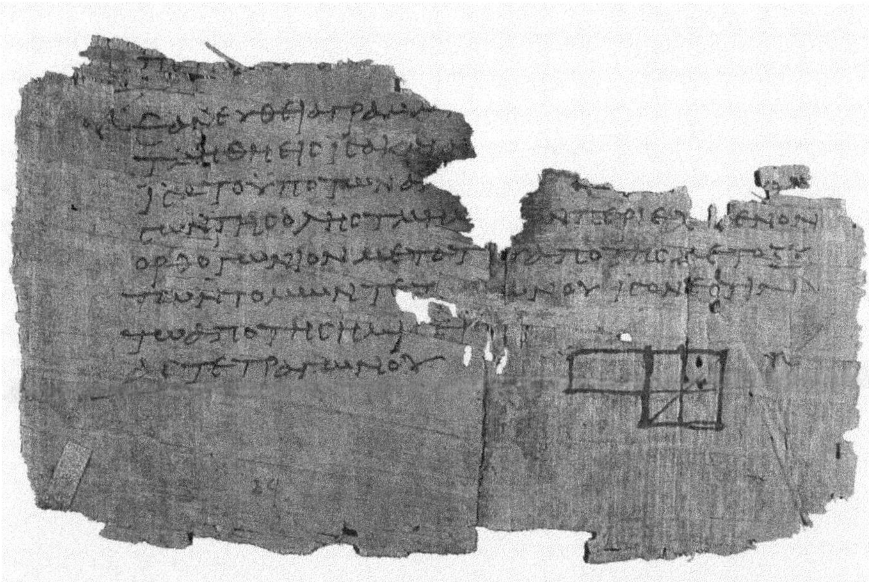

FIGURE 0.4. A fragment of Euclid's *Elements*, from Egypt, *Papyrus Oxyrhynchus* 29, *c*. 75–125 CE. University of Pennsylvania Museum. Photo by Fine Art Images / Heritage Images / Getty Images.

When thinking about ancient mathematics, it is important to keep in mind just how sparse these texts are. Texts are material things: they deteriorate, become damaged, or are lost. The mere fact that Chinese texts were written on silk, wooden boards, and bamboo strips has had much do with their survival. The upshot is that one must be very cautious when it comes to pronouncements about what, say, Pythagoras knew or did not know, or what he proved or did not prove. All accounts of Pythagoras, specifically, are second- and third-hand, typically from the mouths of warm admirers or determined opponents, and mostly from sources written a century or much more after Pythagoras lived. This is also a problem for historiography, for in the absence of texts, how are the gaps to be filled? Is it fair to assume that the thinking of the ancient Egyptians or earliest Greeks, whose works are lost, nevertheless was based on thinking like our own? The situation, strangely enough, is slightly better for Mesopotamian mathematics for the simple reason that their texts were written on clay—and a library, like the great library in Nineveh, when it burned down, left its clay tablets baked and transformed into solid, nearly indestructible bricks! The fact that so many of these tablets remain—literally thousands—is another reason why the case of Mesopotamian mathematics is such an important one. Nevertheless, in reading the cuneiform tablets that survive, it is still necessary to piece together their meanings indirectly, something that can largely be avoided in later periods of history given the direct evidence of letters and texts written by authors explaining their meaning or motives.

As for the *Elements*, it was not the only book of "elements" in antiquity. Proclus speaks of a book of elements, for example, by Hippocrates of Chios (see Chapter 3) and another by a certain Leon. Euclid's work, however, was incontrovertibly the most complete and most influential book of elements in its own time and afterwards. It was not an original work, that is, it was not meant to showcase a series of new results formulated and proved by Euclid himself. The *Elements* was rather a carefully wrought anthology of previously known results (Ptolemy's *Almagest*, incidentally, is similar in this way with respect to astronomy). *How* Euclid's anthology was arranged and what its purpose was cannot be taken for granted.

The first problem concerns the name, *Elements*, or *Stoixeia* in Greek. What does Euclid mean by this? Again, Proclus is helpful here. Writing on the purpose of Euclid's *Elements*, he says:

> If we start from the elements, we shall be able to understand the other parts of this science; without the elements we cannot grasp its complexity, and the learning of the rest will be beyond us. The theorems that are simplest and most fundamental and nearest to first principles are assembled here in a suitable order, and the demonstrations of other propositions take them as the most clearly known and proceed from them. (Morrow [1970] 1992: 58–9)

In the same place, Proclus is very careful to distinguish between being *elementary* and what might be called *elemental* (note here that the words "elementary" and "element" are as close in Greek as they are in English). He says, "'Elementary' propositions are those that are simple and elegant and have a variety of applications but do not rank as elements because the knowledge of them is not pertinent to the whole of the science" (p. 59). Proclus himself is well aware that it is not always easy to say exactly what should be included as an "element," yet he is clear that what one looks for is something elemental in the sense of being fundamental, "close to first principles," forming the ground on which the science can be built.

Just to make this last point clear, it is worth recalling another of the great works of Greek mathematics, the eight books of Apollonius' *Conics*, which is famous for its thoroughness, complexity, and difficulty (see Fried and Unguru 2001). In a letter introducing the first book of his *Conics*, Apollonius says to Eudemus, to whom the letter is addressed, that the first four books have the character of a course in the *elements*. Eudemus, we gather, was a fellow mathematician with whom Apollonius obviously discussed his work and who was clearly interested in that work. When he communicates his elements to

Eudemus, Apollonius instructs him to acquaint *only* those "worthy of sharing in such things" with the work. This suggests that for Apollonius, and presumably for Euclid, a collection of elements was not necessarily for beginners. On the contrary, it was a work of fundamentals meant for initiates—educational, yes, but educational in a higher sense.

Proclus regarded Euclid's achievement in another *more* elemental sense. Elements are also the building blocks of the world, as in the periodic table of chemical elements. The last book of Euclid's *Elements*, Book XIII, concerns the construction of the five regular solids and the relationships among them. Proclus emphasized the point made in Plato's *Timaeus*, that four of those solids represent the "cosmic bodies"—the elements of the world: the tetrahedron is fire; the octahedron is air; the cube is earth; the icosahedron is water; the dodecahedron, represents the aether, literally the quintessence, the fifth element. Proclus says that while Euclid's *Elements* directed the learner to first principles and the understanding of the science as a whole, its main goal was actually the investigation of these cosmic figures, the five regular solids. He says this no less than four times in his commentary! In light of the tremendous amount of *other* things that we find important in the *Elements*, this may seem like an incredible claim. Yet, it is also precisely the kind of claim that raises the question of what is it that makes elements, elements.

One way to begin answering this question is to consider what Euclid includes in his *Elements*. Some of it is close to a high school geometry course, as one might expect, but by no means all of it—and the differences can be surprising and puzzling. The work contains thirteen books, all leading up as noted to a final chapter devoted to geometrical constructions using straightedge and compass alone, of the five perfect, regular solids. Book I launching this journey is on first sight the closest to a standard geometry course. Proclus focused on this book as the exclusive subject of his commentary. It opens with a set of definitions, postulates, and what are called common notions. These are the hallmarks of a Euclidean approach, though it should be said postulates and axioms are only found in Book I.

The axioms or common notions are statements of a very general character such as "the whole is greater than the parts," and "when equals are added to equals the wholes are equal." The postulates are fundamental geometrical demands—the word for postulate in Greek, *aithma*, means "a demand." The postulates tell us, for example, that a line may be drawn from any point to any other, that all right angles are equal, and a circle can be drawn with any center and radius. One of the postulates stands apart in lacking this same simplicity and straightforwardness. It is the famous fifth postulate which states:

> If a straight line falling on two straight lines make the interior angles on the same side less than two right angles, the two straight lines, if produced

indefinitely, meet on that side on which are the angles less than the two right angles.

Euclid uses this postulate to prove that if any two parallel lines are cut by a transversal, the alternating angles are equal, and from that that the sum of the angles of a triangle must equal two right angles. This is a postulate that many mathematicians ever since Euclid have taken to be a theorem, that it should be something provable from the axioms and postulates of geometry, and indeed, many attempts have been made throughout the history of mathematics to prove the parallel postulate as a theorem. The famous French mathematician, Joseph Louis Lagrange, was among them, and as late as 1806 believed he had proved the postulate as a necessary result of Euclidean geometry (see Volume 5, and Grabiner 2009: "Why did Lagrange 'Prove' the Parallel Postulate?"). Later in this series, the reader will find that investigations of geometry without this postulate or with its denial led to new kinds of non-Euclidean geometries (Volume 5). In the twentieth century, Albert Einstein's theory of relativity led to experimental verification that space is actually curved, i.e. it is non-Euclidean (Volume 6). Euclid, however, thought otherwise, or so he postulated. As for the propositions in Book I, the first proposition of the forty-eight that make up the book shows how an equilateral triangle can be constructed on a given line, a proposition which is then used to show how a line may be constructed from any given point having a length equal to a given length. These constructions are the first in a series that can be executed with compass and straightedge. Such constructions are well known and often thought to be characteristic of Greek mathematics. But this is only partly true. One finds other kinds of constructions in Greek mathematics that cannot be formulated in terms of moves with compass and straightedge. For example, there is a kind of construction called *neusis* in which a line is drawn from a point so that a given length is cut off between two curves (usually two lines or a line and a circle). Archimedes uses this kind of construction to trisect an angle, a construction impossible to carry out with compass and straightedge.

Book I also contains the usual congruence theorems for triangles, areas of figures, and the "Pythagorean theorem," (I.47) with its iconic diagram (Figure 0.5).

Book II is the book most difficult to describe and was, as we said, the focus of a historiographical debate concerning the aptness of modern conceptions—in this case, algebra—in interpreting ancient texts. To get a feel for the content of the book and a hint of what was at stake in the debate, consider proposition 5. It states: "If a straight line be cut into equal and unequal segments, the rectangle contained by the unequal segments of the whole together with the square on the straight line between the points of section is equal to the square

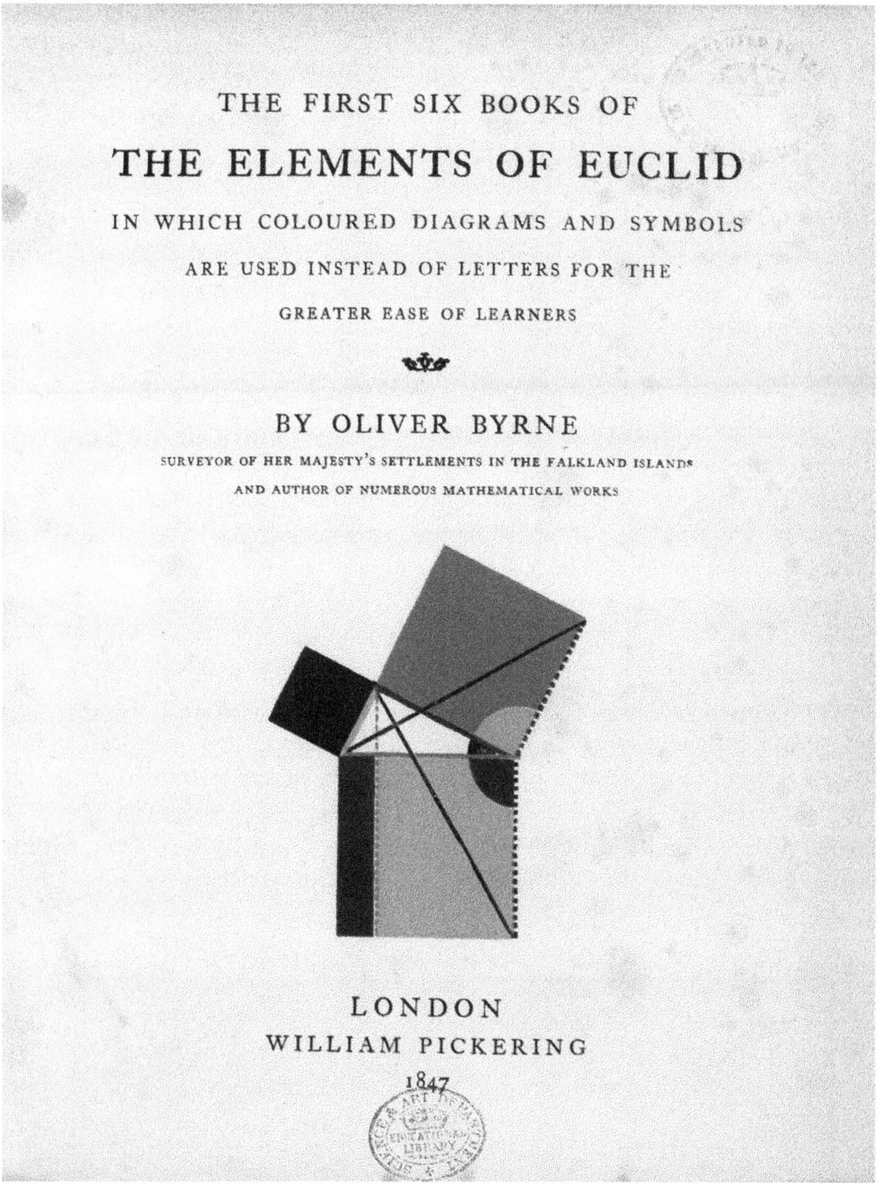

FIGURE 0.5. Iconic diagram of the Euclidean proof of the Pythagorean theorem on the cover of the 19th-century "Byrne Euclid." Photo by Science & Society Picture Library (SSPL) / Getty Images.

on the half." Thus, if the line is AB and it is divided in half at C and at some other point D, it needs to be shown that the rectangle contained by AD and DB together with the square on CD is equal to the square on CB.

Euclid's diagram[7] itself essentially provides the proof (Diagram 0.3).

As Euclid tells us, the rectangle contained by AD and DB is equal to the shaded figure LDBFG; then, adding the square LG which is the same as that built on CD, he has the result. This shifting of areas like tiles is reminiscent of other arguments in Book I, in particular, proposition I.35, that: "Parallelograms on the same base and in the same parallels are equal to one another." It, too, is about the equality of figures not having the same shape. There is in fact a kind of nexus beginning with I.35 continuing to problems concerned with constructing figures of different shapes equal to a given figure (e.g., I.45, to construct a parallelogram with a given angle equal to a given figure) to the Pythagorean theorem concerned with two figures of the same shape which together are equal to a third of the same shape, to these propositions in Book II: significantly, the last proposition of Book II, "To construct a square equal to a given rectilineal figure," returns to the line of thought begun in proposition I.45.

Interestingly enough, surviving texts on ritual practice in late Vedic India, from at least the early first millennium BCE, reflect a similar preoccupation with area-preserving rearrangement of geometric shapes, in this case for the construction of sacrificial altars of baked brick. The alteration of a square into an oblong rectangle, or trapezoid or circle, with equal area has a cosmic and spiritual significance as well as a demonstrably accurate prescribed geometric procedure.

But if one departs from a way of thinking guided by this geometrically appealing rearrangements of shapes, and if one thinks algebraically, one can see

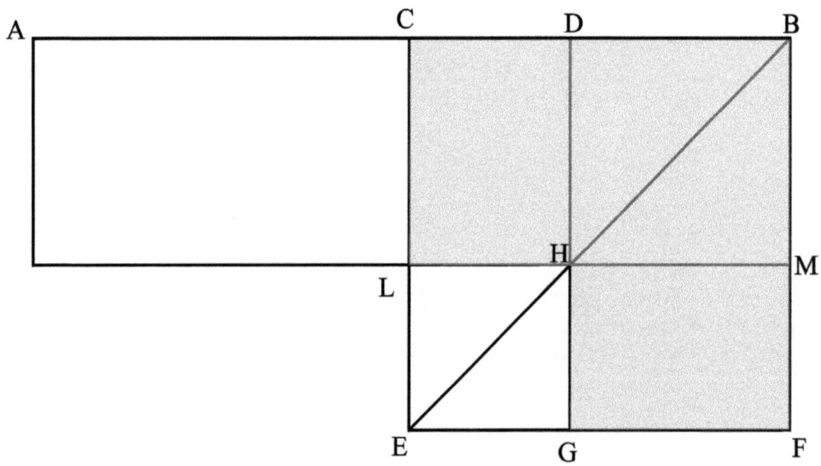

DIAGRAM 0.3. Euclid's diagram for *Elements*, II.5.

something quite different in a proposition such as II.5. For example, if the geometric problem of showing the equivalence of the given areas is translated into algebraic terms, this begins to look very different: Let $AC=CB=a$ and let $CD=x$. Then $AD=a+x$ and $DB=a-x$, so that the rectangle contained by AD and DB has an area equal to $(a+x)(a-x)$. Thus, the proposition says $(a+x)(a-x)+x^2=a^2$ or $(a+x)(a-x)=a^2-x^2$. Other basic algebraic identities can be read into other propositions in Book II: proposition II.1, on this reading, shows $a(x+y+z)=ax+ay+az$; proposition II.4 shows $(a+x)^2=a^2+2ax+x^2$, and so on. This view of Book II as presenting in a geometric format rules of algebra and means of solving algebraic equations—for II.4 to II.6 can be used to solve quadratic equations—is what became known as the "geometric algebra" interpretation.

In later times, this book *was* used to justify algebraic procedures by geometric demonstrations, especially in the Islamic period. The historian Leo Corry (Corry 2013) followed this development up through the Middle Ages, showing how the character of Book II transformed from one that was decisively geometric into one that was arithmetic and algebraic. The power of algebra, as Descartes and others recognized, is its ability to reveal abstract patterns in widely diverse settings: if one knows algebra, it can be found almost everywhere. But is this enough to say that Euclid was also thinking in these terms, albeit with the clumsy language of geometry?

The historian must approach this question from the point of view of Euclid, the texts he wrote, and those available to him. This is basic to history. It was on this ground that Unguru, mentioned above, saw a need to rewrite the history of mathematics, which he saw as driven then by the logic of our own, admittedly powerful, mathematics. One challenge to this has been the discovery in Babylonian texts of similar geometrical reasoning for problems that were certainly arithmetical—and in a qualified way, algebraic (see Høyrup 2002)—as will be seen in the first part of Chapter 3 below. Yet, even with that, one cannot assume that Euclid was driven by the same logic as those in the Near East rather than the kind of nexus of ideas alluded to above.

Before leaving this issue, some remarks about Chinese mathematics are in order. For here, too, one finds reasoning based on the rearrangement of figures to produce equal areas. And here, too, it is possible to give algebraic interpretations, though in this instance a stronger case might be made for such an interpretation, or at least for the mix arithmetic and geometrical reasoning.[8] In some of these instances, the geometrical character of the reasoning is unmistakable and even assume to some degree the form of visual proofs-without-words. For example, Liu Hui, the third-century CE commentator on one of the classic Chinese mathematical texts, the *Jiuzhang suanshu*, or *Nine Chapters on the Mathematical Art* proves a proposition which is the Chinese equivalent of the Pythagorean theorem by dissecting and rearranging the areas of the squares on the two sides of the right triangle lying "outside" of the hypotenuse square and then moves them "inside" to form the square on the hypotenuse (Figure 0.6).

The Chinese, like the Pythagoreans, proved many propositions involving right triangles: indeed, Chapter 9 of the *Nine Chapters* is entirely devoted to problems that also occur in Euclid's geometry, particularly those in Book IV of the *Elements* concerned with inscribing and circumscribing figures. On the other hand, we also find problems, such as the following, which are unmistakably algebraic in intent, but solved in terms of geometrical figures:

> Suppose there is a door whose height and width are not known, and a pole whose length is unknown. The horizontal [width of the door] falls short *bu chu*] by 4 *chi*, and the vertical [height of the door] falls short by 2 *chi* [of the length of the pole forming the diagonal], but the diagonal fits exactly [*shi chu*]. The question: How much are the height, width, and diagonal of the door?
>
> The answer: The width is 6 *chi*; The height is 8 *chi*; The diagonal is 1 *zhang*. (Dauben, Guo, and Xu 2013: vol. 3, xxx)

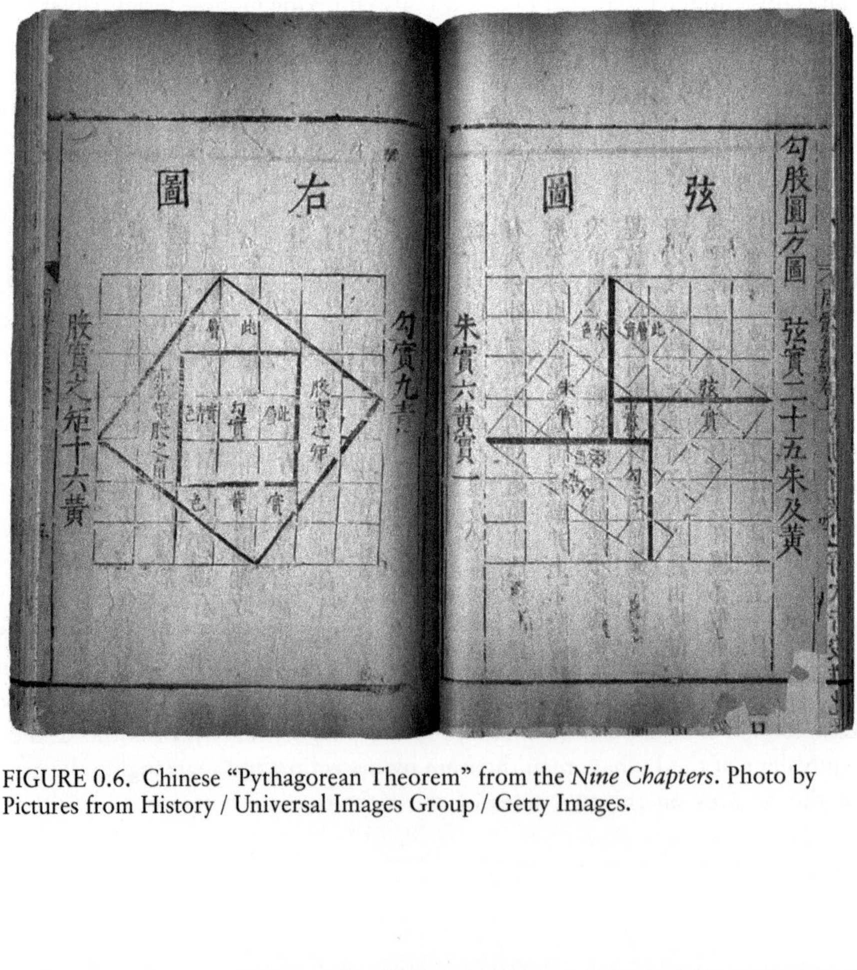

FIGURE 0.6. Chinese "Pythagorean Theorem" from the *Nine Chapters*. Photo by Pictures from History / Universal Images Group / Getty Images.

In algebraic terms, we begin by letting a, b, and c represent the base, altitude, and hypotenuse of the right triangle. Then we have, according to the problem, $c\text{-}a=4$ and $c\text{-}b=2$. The geometrical presentation of the solution, however, can be seen in a sequence of diagrams from a Song dynasty work by Yang Hui (1261). The key diagram contains three squares, an outer square whose side is c, a square drawn from the lower-left-hand corner whose side is b, and another form the upper-right-hand corner whose side is a (Diagram 0.4). The rectangles in the upper left and lower right, then, have area $(c-a)(c-b)$, while the central square has area $(a+b-c)^2$.

The result, then, hinges on the fact that $2(c-a)(c-b)=(a+b-c)^2$, and $a=(a+b-c)+(c-b)$. The method for solving this problem says that the square root of twice the excesses of the diagonal over the width and height of the door, plus the excess of the diagonal over the height, gives the width of the door: $a=\sqrt{2(c-b)(c-a)}+(c-b)$. Since we know, though, that $c\text{-}a$ and $c\text{-}b$ are 4 and 2, respectively, we find $a=6$, $b=8$, and $c=10$ (or 1 *zhang*).

But let us return to Euclid. We shall not go into any detail about Book III except to say that it concerns circles and their properties. This is not because

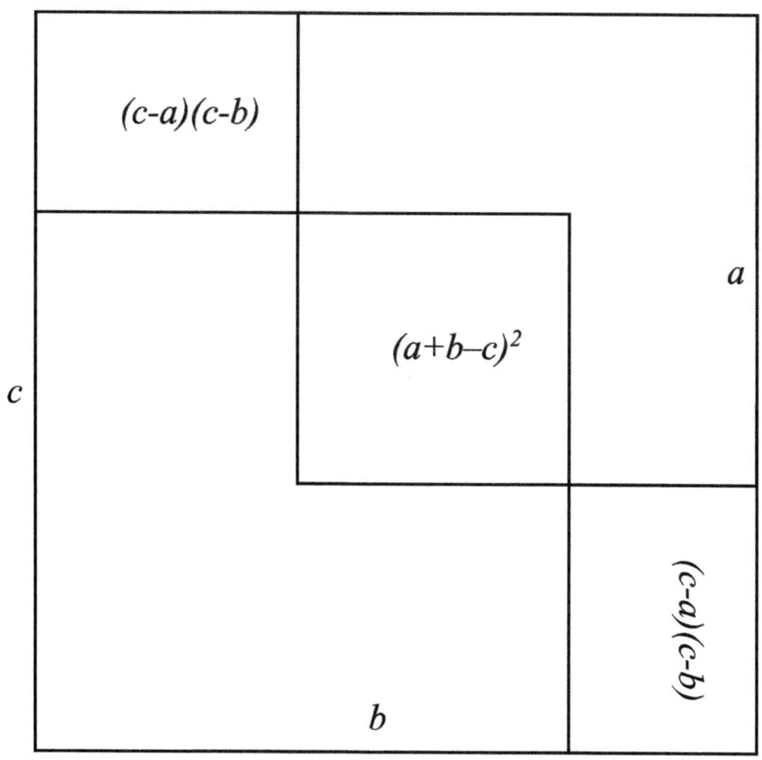

DIAGRAM 0.4. Yang Hui's solution to the "door problem."

there is no more to say about it: in a longer discussion exclusively dedicated to Euclid we would show, for example, the fascinating way in which the end of the book harkens back to Book II and, perhaps, to some of its difficulties. Book IV, as we mentioned above, continues the work of Book III in focusing on problems of circumscribing figures and inscribing figures in circles. By end of Book IV, one has learned how to inscribe a triangle, square, pentagon, hexagon, and a fifteen-sided figure in a circle, which is the same as constructing the figures themselves. How to construct the regular seven-sided heptagon is not shown. That became a celebrated problem and was finally shown to be impossible by the great Carl Friedrich Gauss at the beginning of the nineteenth century.

The next book, Book V, stands out from the other books of the *Elements*, certainly from Books I-IV, in its complete generality. Its subject is ratio and proportions among magnitudes. The book never defines exactly what a magnitude is, except, perhaps, to say only that a magnitude is that which can put into a ratio with another. Ratio itself is not very well defined either. It is "a sort of relation with respect to size between two magnitudes of the same kind" (Book V, def. 3)—thus, a length with a length, an area with an area, a weight with a weight, a rectilinear angle with a rectilinear angle. Two magnitudes have a ratio when each can be multiplied so as to exceed the other. It is not a trivial characterization: already Proclus described the so-called "horned angle" between a circle and its tangent and for which this property does not hold.

Be that as it may, the most fascinating and maybe most significant part of Book V is its definition of the sameness of ratio, that is, of proportion. This would be a simple matter, if given any two magnitudes, say two lengths, A and B, one could say that A is 3/5B, so that A:B is the same as 3:5. Then any C:D would be the same ratio as A:B if also C:D is the same ratio as 3:5. But it turns out that there are magnitudes which are not the same ratio as some number m to another number n. For example, if the magnitudes happen to be the diagonal of a square D and its side S, there are *no* numbers m and n such that D:S is the same as m:n. Nevertheless, for such "incommensurable magnitudes" (see Chapters 3 and 4 and below), Euclid's definition works just as well as it does for any other magnitudes. This definition—and probably the rest of Book V—was most likely due to a thinker named Eudoxus who lived during Plato's time. It is a good example of how Euclid's work was a kind of carefully organized collection of known work rather than a presentation of his own original results, as mentioned above.

Book VI applies the theory of ratio and proportion set out in Book V and develops the idea of similarity. Books I–VI, then, contain the material of a typical high school geometry course. Thus, for example, editions of Euclid designed to be textbooks, particularly John Playfair's famous *Elements of Geometry* (first published in 1795), contained only these books. Likewise, when Matteo Ricci set out to translate Euclid's *Elements* into Chinese with his colleague Xu Guangqi, they only published the first six books in their edition of 1607. On this basis then, one might reasonably think that *The Elements* is short for *The Elements of Geometry*, as in Playfair's textbook. But coming to Books VII–IX, one is surprised to discover that these books are all about numbers, what we might call "number theory."

Books VII–IX contain what is today called the Euclidean algorithm for finding the greatest common divisor of two numbers (VII.1–2), the proof of the infinity of primes (IX.20), and the fundamental theorem of arithmetic (not as one theorem, but garnered from VII.30–32 and IX.14). The language of these books is that of "measuring" (*kata-metrein*, in Greek, as in the word "metric"). It is the language of multitudes where one measures by counting. This is reflected in the very definition of a number as a multitude composed of units (VII, def. 2), a definition which excludes fractions from being numbers at all! As if to underline how these books diverge from Books I-VI, a different definition of proportion is given here, one tailored for numbers: *numbers* A, B, C, and D are in proportion, $A{:}B$ as $C{:}D$, whenever there are numbers M and N such that when $MA=NB$, also $MC=ND$. And we recall that such numbers M and N may not always be found if A, B, C, and D were *magnitudes*.

In considering this sharp delineation between numbers and magnitudes in the *Elements*, one should recall the Pythagorean division of the mathematical sciences described above. According to that scheme, Books VII–IX belong among the sciences along the branch concerned with multitudes as opposed to magnitudes. It may be, too, that these books were inspired or at least closely connected to music, the other science on that branch. Many of the propositions in these books, particularly from Books VII and VIII, in fact appear one way or another in Euclid's *Sectio Canonis*, Euclid's book on music theory (see Barker 1989: 194).

These books, then, force us to ask what exactly are the "elements" contained in the *Elements* the elements of? Can we still say, *geometry*? Books VII–IX are certainly not about geometry, yet they are in some ways still geometrical. The language of some of the terms is geometrical, for example, there are references to "plane numbers," "solid numbers," "square numbers," "cubes," "similar plane and solid numbers." More strikingly, the diagrams always represent numbers as line segments that can be added or subtracted as line segments. And yet, while plane numbers are those produced by *multiplying* two numbers, no rectangle in the *Elements* can be produced by multiplying its sides: rather, it is always that a rectangle is *contained* by its sides (literally, "under its sides" as if the sides are a frame for it).

Later in the history of mathematics, Descartes, in Book I of his *Géométrie*, will say that the Greek's hesitation (the word he uses is "scruple") to use terms from arithmetic in geometry kept them for making the kind of algebraic discoveries for which Descartes was famous.[9] But that scruple they had nevertheless. The discovery of incommensurable magnitudes played some part in showing that arithmetic was incomplete and could not deliver the full generality of geometry. Whatever the reason, the so-called "arithmetic" books of the *Elements*, Books VII, VIII, and IX, are evidence that Euclid's book is more than just the elements of geometry alone.

The next books move back to geometry though taking in to some extent the material in arithmetic books. Book X is by far the longest of the thirteen books, and it is the book that finally takes the theory of incommensurable magnitudes by the horns. These incommensurable magnitudes, unlike numbers, can have no common measure, or, as we described them above, magnitudes A and B for which there are no numbers m and n with $mA=nB$. The theory arises out of arithmetical ideas such as common measure which were developed in the previous three books. Book X thus begins with a process very similar to the "Euclidean algorithm" with which Book VII began; however, it continues in a geometric vein. It is in this way a kind of bridge, albeit one of unlikeness, between arithmetic and geometry.

The discovery of these "incommensurable magnitudes" in the Classical world, traditionally attributed to the Pythagoreans (see Chapter 4), was of immense importance. It was a discovery that not only preoccupied people directly engaged in mathematical investigations but also mathematically-minded philosophers like Plato. We mentioned above the Platonic dialogue *Meno* and its famous mathematics lesson. That lesson that Socrates gives to a slave boy involved finding the side of a square who area was twice that of a given square. After initial attempts, for example that the side is twice the side of the given square, Socrates leads the slave boy to see that the side required is the diagonal of the original square. The lesson was impressive because the answer was not a multiple of the side that one could say (at one stage, Socrates tells the boy, "If you cannot say which line, just point"): the diagonal of a square (a length we would write today as T2 for a unit square) and its side is a simple example of two magnitudes that are incommensurable, striking *because* it is so simple. No line that can measure the side can measure the diagonal. These things were of deep concern and deep interest to the Greeks, a reason for caution in treating magnitudes in terms of numbers. Today, this would deserve Descartes' term, a mere "scruple," but not then.

In this the Greeks were fundamentally different in their approach from the Chinese. The Chinese knew that for a unit square, the diagonal could be approximated, and they were content to compute the square root of any given number to whatever degree of accuracy might be wished, and then stop computing. And it was not only about the diagonal and a square and its side that they had no qualms. When it came to circumference of a circle with a unit diameter, a length that we now indicate with the Greek letter π, the Chinese knew that no whole numbers m and n could be found so that $\pi=m/n$, but they accepted 3 as a general, useful approximation for most everyday purposes, though mathematicians offered more precise values such as 22/7, 157/50, or 355/113, for example. The particular case of π is, of course, important and celebrated, so we shall take up in more detail below. But let us come back to our survey of the *Elements*.

Books XI to XIII returns solidly to geometry—indeed, to solid geometry! Book XIII, as already mentioned, concerns the five regular solids. Book XI demonstrates general facts concerning the geometry of figures in space—lines and planes in space, solid angles, pyramids, and parallelepipeds, among other things. Book XII investigates ratios and proportions connected to solid figures as well as their volumes, for example that a cone is a third of a cylinder having the same height and base (XII.10). But the distinctive character of Book XII lies in the procedure which dominates the book, what we now call "the method of exhaustion." It is worth taking a closer look at this method since it has been seen as an anticipation of integral calculus.

As is typical in Greek mathematics, "the method of exhaustion" is not stated as a method, but simply used. To get an idea of how it works, consider proposition 2 of Book XII. It is not about a solid figure, but it very clearly exemplifies the technique.

Proposition 1 of Book XII proves that if similar polygons are inscribed in circles then their areas have the same ratio as that of the squares on the diameters of the circumcircles. So, in Diagram 0.5, the inscribed polygons have the same ratio as the squares BD and FH. What proposition 2 proves is that this is *also* true about circles themselves, that is, instead of the polygons, the circles ABCD and EFGH themselves are in the same ratio as the squares on BD and FH.

To prove this, Euclid assumes that the proportion does *not* hold, so that instead of EFGH there is some figure S less than or greater than EFGH with:

circle *ABCD*: *S*::square *BD*: square *FH*.

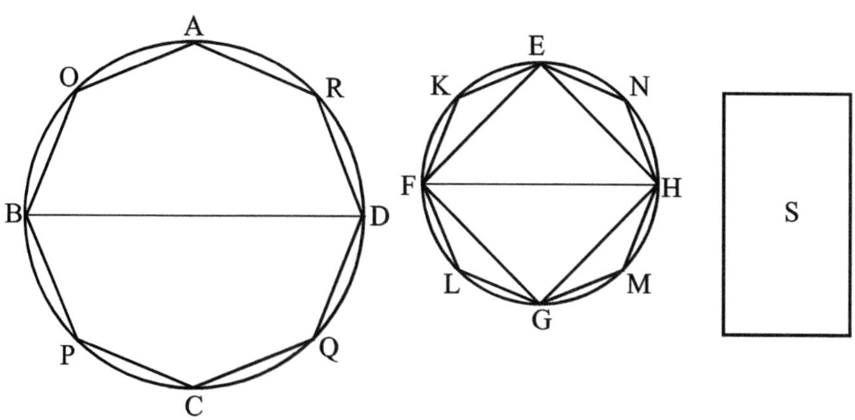

DIAGRAM 0.5. Euclid's diagram for *Elements*, XII.1.

Euclid first considers the case where S is less than EFGH. He then constructs polygons inscribed in EFGH with greater and greater number of sides, and similar figures in the circle ABCD. By the first proposition, these pairs of similar polygons (that is, their areas) all have the same ratio as the squares on BD and FH. Euclid shows that with a sufficient number of sides, a polygon can be inscribed in the circle EFGH that is *greater* in area than *S*. But this is impossible, since circle *ABCD* must be greater than the similar polygon inside of it, so that the ratio of ABCD to S must be greater than that of the two polygons, that is, greater than ratio between the square on BD and that and FH. From this contradiction, Euclid concludes that *S* cannot be *less* than the circle *EFGH*. By a similar argument, he shows it cannot be *greater* either. Therefore, *S* must be *equal* to the circle *EFGH*.

The way in which the area of the circle is "exhausted" by the successively inscribed polygons until the difference between the circle and the polygon is less than that between the circle and S is reminiscent of a limiting processes. But, like the algebraic interpretation of Book II, this is true chiefly because one may have a limiting process already in mind from previous experience with the infinitesimal calculus pioneered by Newton and Leibniz (Volume 3). As historian of the calculus Carl Boyer ([1949] 1959) points out, the term "method of exhaustion" dates from the seventeenth century, "when mathematicians somewhat ambiguously and uncritically employed the term indifferently to designate both the ancient Greek procedure and their own new methods which led immediately to the calculus and which truly 'exhausted' the magnitudes" (p. 34). But Boyer goes on to stress that while the type of argument is similar to those from the integral calculus, "it does not represent the *point of view* involved in the passage to the limit" (p. 35). One might argue that, as with the case of Book II, vis-à-vis algebra, the development of the calculus involves a radical reinterpretation of some of these ancient results. That, however, is something completely different from claiming that the calculus was somehow already in Euclid's sights.

HISTORIOGRAPHICAL REFLECTIONS REGARDING π

Beyond the sophistication of Euclid's proof and its importance when the method of exhaustion is brought back and reinterpreted in the seventeenth century, as Boyer's remarks make clear, proposition XII.2 raises a question about that most ubiquitous symbol in all of mathematics: π. The formulation of proposition XII.2 is characteristic of Greek mathematics generally: in speaking of areas or volumes, a relation is given rather than a precise measure or formula for numerically determining the area. Thus, for example, everyone learns in school that the area of a parallelogram is base×height (so that the area of a triangle is ½ base×height). But instead of such a formula, in the *Elements*, what one finds is that all parallelograms between parallels (i.e., having the same height) and on the same base are equal to one another. From that, one can go on to conclude that those parallelograms are equal, in particular, to the *rectangle* between the parallels and on that same base. But Euclid does not make this argument: I.35, the proposition about the parallelograms, concerns the equality of non-congruent figures, and not a formula for area.

Proposition XII.2, similarly, proves that the areas of circles are in the same ratio as the squares on their diameters, or, what is equivalent, the squares on their radii—but Euclid does not say what the "constant of proportionality" is, as might be expected after such a statement. Archimedes, when he considers the circle, also begins with a relation. In his work *Measurement of the Circle*, he first shows, using a method along the lines of XII.2, that the area of a circle is equal to a right triangle, one of whose legs is equal to the radius of the circle and the other to the circumference of the circle: whatever the area of the circle is, it is the same as that of a certain triangle. Then he shows (Proposition 3) that the circumference of a circle is less than 3 times the diameter and 1/7th more but greater than 3 times the diameter and 10/71th more. Considering the exact ratio between the circumference and the diameter *C/d* (as mathematicians did until early in the eighteenth century, when it began to be viewed as a *bona fide* number, although one with a value that could only be approximated), today one would say that $3.1408 \approx \frac{223}{71} = 3\frac{10}{11} < \pi < 3\frac{1}{7} = \frac{22}{7} \approx 3.1429$. However, Archimedes did not express C/d as a number serving as a constant of proportionality. A number, for the Greeks, was a whole number, as we have seen. So the *number* is 3; the circumference exceeds 3 times the diameter by less or more of a certain number of parts of the diameter. It is with these latter numbers that Archimedes is concerned: 1 out of 7 parts and 10 out of 71 parts.

Our remarks about China and incommensurable magnitudes brought out the fact that other cultures have asked about the ratio C/d. The ancient Sanskrit ritual geometry texts mentioned above likewise imply in their manipulations of circular areas various approximate values of C/d, including $(26/15)^2$ and 25/8. The number 3 appears as an approximate value for C/d in the Bible. In 1 Kings 7:23 (and in 2 Chronicles 4), the text describes a kind of huge brass pool, a "molten sea," meant for priestly ablution: "[Solomon] made the molten sea of ten cubits from brim to brim . . . and a line of thirty cubits did compass it round about." In a popular book on the history of π (Beckmann 1971), the conclusion is then drawn that "the biblical value of π is 30/10=3" (p. 15). The author, Petr Beckmann, an electrical engineer, then goes on to remark that: "At that time [around 550 BCE when the Bible was edited], π was already known to a considerably better accuracy, but evidently not to the editors of the Bible" (ibid.).

Suggesting that the editors of the Bible were interested in a number representing the constant of proportionality for the relation between the circumference of a circle and its diameter is something that should be questioned in a book such as this concerned with the cultural history of ancient mathematics. That workers in the court of Solomon had a rule of thumb for building a circular structure is reasonable enough, but was it any more than that? Even in the case of Archimedes, who did not seem to have any practical activity in mind and certainly *was* concerned with the mathematical relations connected to the measurement of a circle, to say that he was groping towards a number like π and could only say for sure that it was between $\frac{223}{71}$ and $\frac{22}{7}$ is no less problematic. This, too, can be said about the Egyptians, whose mathematics, like that of the Babylonians in the Near East, has been celebrated as a great achievement of the ancient world.

A sign of that achievement is the putative calculation of π ascribed to the Egyptians. Indeed, it is chiefly that which Beckmann has in mind when he says that when the Bible was being edited, "π was already known to a considerably better accuracy." This calculation is found in a beautifully preserved papyrus known today as the Rhind Papyrus, named after Alexander Henry Rhind, a Scottish collector and archaeologist who purchased the papyrus in Luxor in the late 1850s and brought it back to Britain. The Papyrus is a long scroll, measuring 513 cm in length and 32 cm in width (for this and the details which follow, see Robins and Shute 1987). Unlike more formal inscriptions, which are generally written in hieroglyphics, the Rhind Papyrus was written in the cursive form of writing called hieratic (see Figure 0.7). The scroll was found, apparently, near the memorial temple of Ramses II at Thebes. It is the work of a copyist named Ahmose, living in the mid-sixteenth century BCE; he, in turn, was copying the work of an unnamed scribe (following Robins and Shute in this terminology) living some 300 years earlier. (Whether or not a name like Ahmose is attached to a mathematical work is itself an important characteristic of ancient mathematics; see Chapters 2 and 3.)

The papyrus contains 84 mathematical items, problems, and procedures, of which the series 41-43, 48, and 50 contain the problems of interest here. As is typical of problems in Egyptian papyri, these are given in terms of a practical problem with specific numbers, and a specific procedure is given for finding an answer. For example, problem 41 asks to find the volume of a circular (i.e., cylindrical) granary, or rather how much grain the granary can hold, when the cylinder has a base with diameter of 9 and a height of 10. In fact, the whole series of problems through problem 47 is devoted to grain in granaries, not only cylindrical ones—*this* seems to have been the main interest of the scribe.

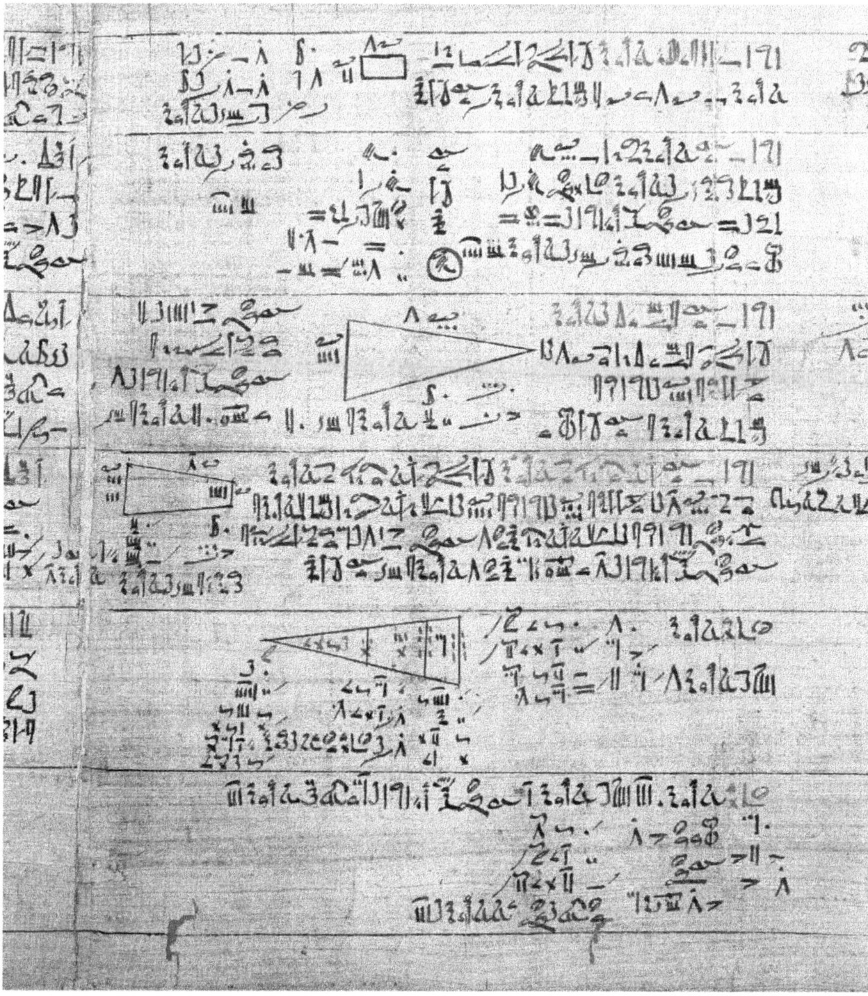

FIGURE 0.7. A portion of the Rhind Papyrus, in hieratic script, *ca.* 1650 BCE. Photo by DEA Picture Library / De Agostini / Getty Images.

Problem 50, the second problem in a new series about areas, asks specifically to find the area of a circle:

> Example of producing [the area of] a round field of diameter of 9 *khet*. What is the reckoning . . . of its area?
> Take away 1/9 of it (the diameter), namely 1; the remainder is 8. Multiply 8 times 8; it makes 64. [Therefore] the amount of it in area is 64 *setjat*. (Trans. from Clagett 1999: 162–3)

The same procedure of removing 1/9 the diameter and squaring the remainder to find the area is used in problems 41-43. It is this which some have taken to reveal "the Egyptian calculation of π." For comparing this procedure to the familiar formula for the area of a circle of radius r, we find:

$$\text{Modern Area(circle)}=\pi r^2=\text{Egyptian Area(circle)}=(d-(1/9)d)^2$$

And, therefore, so the argument goes, the Egyptians *must have known that* $\pi r^2=((8/9)(2r))^2$ and thus $\pi r^2=4(64/81)r^2$, i.e. $\pi=256/81$. So the "Egyptian value of π" was $256/81=3.16049\ldots$ The historian of Egyptian mathematics, Annette Imhausen, calls this the "Myth of the Egyptian π," the first of several such myths she discusses in connection with Egyptian mathematics (Imhausen 2010). Referring to problem 50 and the procedure of removing 1/9 the diameter and squaring to find the circle area, she says that, "While this may be a sound way to establish how accurate the Egyptian procedure was, it remains historiographically incorrect to state that the Egyptians used an approximation of π which was 256/81. The Egyptians did not use π" (p. 789). She then points out that it was *not this* but the fraction 1/9 which embodied "the ancient Egyptian method of mathematizing circles." In other words, the numbers appearing in these Egyptian texts were *part of a procedure* for working through a problem requiring the area of a circle, not a constant representing the ratio C/d.

The myth of the "Egyptian π" provides one more cautionary lesson about how careful one must be in considering the *cultural* history of mathematics. The point, at bottom, is the same as it was in regard to Archimedes. It is not that 256/81 is not a good approximation for π, for it is: the problem is rather the blithe assumption made by Beckmann and others that the scribe must have been *looking* for a number and a relation such as Area(of a circle)$=\pi r^2$. The importance of π in modern mathematics is so overwhelming—even at an early age, students learn to approximate it as 3.1415—that it is easy to assume that anyone at another place or time must have understood it on similar terms as we do. Yet, it must not be forgotten that even thinking of π as a number at all is part of a modern conception of number, as will be clear in Volumes 5 and 6 of this series. Indeed, the idea of using the Greek letter π, the first letter of the Greek word for circumference, *periphereia*, to indicate the ratio circumference: diameter was the inspiration of an early eighteenth-century *Welsh* mathematician, William Jones! And then it took a *Swiss* mathematician, the great polymath Leonard Euler, to popularize its use among mathematicians!

The long discussions we have presented of Euclid's *Elements* and of π show how these things which seem familiar to us become less obvious when we consider them in the light of history, culture, and the nature of mathematical activity. More specifically, from these discussions and throughout this introduction, it should now be clear that in order to understand our mathematical past, we must be mindful that our contemporary understanding of mathematics does not always reflect what earlier mathematicians, or those who were pioneering or applying mathematics in a wide variety of ways, either did or thought they were doing. In short, the challenge for us in this cultural history of mathematics is letting the many cultures of the past speak on their own terms, as strange as those terms may be and as far as they may be from what seems to us undeniably obvious and universal.

CHAPTER ONE

Everyday Numeracy

Mathematics in the Everyday Lives of Citizens in Greco-Roman Times

MERAV HAKLAI

Numeracy is the ability to count, calculate, and measure. It is based on natural abilities of humans, such as, a basic ability to estimate size and small clusters and, perhaps, a certain ability to count (Dehaene 2011). Yet, also, "numeracy is viewed as a *cultural practice* that builds on innate mechanisms for understanding quantities. . . . *cultures* influence mathematical practices through the *belief systems* associated with numeracy, as well as through *tools and artefacts* (e.g., symbol systems) that support numeracy" (Okamoto, Brenner, and Curtis 1999: 611, emphases are mine). This chapter is concerned with these cultural and historically located aspects of ancient numeracy.[1] It focuses chiefly on Greco-Roman civilizations, though it includes some remarks on Mesopotamian and (to a lesser extent) Egyptian cultures. More specifically, this chapter concerns the everyday use of numbers and calculating systems by the general public, that is, by non-professionals.

Both "general public" and "non-professionals" are ambiguous categories. While it is easier to identify as "professionals" high-ranking state officials or persons allocated with complicated tasks, such as, land-surveying, architecture, or financial bookkeeping, it is harder to define "the unskilled." In ancient societies, work was carried out by different individuals, from unskilled laborers to those with practical experience, from those with some expertise to those with specialist knowledge. The definition of "specialist," "experienced," or "unskilled" is flexible, to say the least. It is often difficult to decide whether or not a person qualifies as a professional and it is almost impossible to set clear criteria to distinguish between practitioners and "the generally numerate." A practical approach is to treat as practitioners those who identify themselves as belonging to a profession. This, however, is not without difficulties, especially since, to modern eyes, some ancient jobs may not be easily recognized as professions. And this is all the more so with professions that we consider as requiring professional numerical knowledge. In fact, at least in Greco-Roman antiquity, almost everyone had to deal regularly with numerical notions and often with highly numerate professionals; hence, they would need some level of numeracy to handle their affairs to the best of their interest or to have a close, trust-based relationship with a person who had such knowledge.

The following section gives some notion as to what numeracy might have meant in antiquity by focusing on the representation of numbers in language and body gestures, in calculative technics, in writing methods, and in numerical education. The rest of the chapter focuses on aspects of civic life and economic conduct to reflect on levels of numeracy, that is, who was numerate and why.

REPRESENTING NUMBERS

Orality and Body Gestures

Humans perceive numerical relations intuitively through the natural world around them, starting with their own bodies, first and foremost with their fingers (Dehaene 2011: ch. 4). Human ability to rotate fingers towards each other makes our ten fingers the first, natural abacus (Netz 2002: 324). Calculating with fingers probably sparked the first decimal as well as base 5 and base 20 systems. In archaic Mediterranean, counting and calculating with fingers was a common practice, attested in the Homeric use of the verb *pempathai*, meaning "to count on the five fingers" or "to compute by five" (*Od.* 4.412). Counting on one's fingers was commonplace for fifth-century BCE Athens (Aristophanes, *Wasps* 656), as well as for fourth-century CE Rome (Ambrose, *De Tobia* 7.25). It was an integral part of everyday life also in sophisticated commercial environments. For example, in the first century BCE, Cicero wrote to his friend Atticus about a Senate decree fixing a 1 per cent interest as the legal rate and assumed that Atticus was calculating the financial implication of the decision as he was reading the letter; hence, Cicero fondly says, "if I know your fingers" (*Ad Atticum* 5.21.13). Apparently, in late republican Rome, calculating literally meant using one's fingers.

Finger counting involved much more than, say, using one finger for 1, two fingers for 2, and so on. In the Greco-Roman world, a sophisticated system of finger notation was developed. By the late Roman Republic, a remarkably uniform system of finger symbols was used throughout the Mediterranean and beyond, as is evident from a recently discovered *tessera* from Phanagoria in the Cimmerian Bosporus at the Black Sea (Ostapenko 2016). It continued from antiquity into the Middle Ages in both the Latin West and the Byzantine East (Williams and Williams 1995). No ancient source actually details the complete system that was used, probably because there was no need to describe something everyone knew; but, there are medieval texts from which earlier versions may be inferred, namely the *Romana computatio* (*c.* 688 CE) and Bede's *De temporum ratione* titled *de computo vel loquela digitorum* (*c.* 725 CE). In Rome, finger numbers were an integral part of computation (Apuleius, *Apologia* 89; second century CE) and a person who lacked facility in using his fingers was deemed uneducated (Quintilian, *Institutio Oratoria* 1.10.35; first century CE). Finger numbers provided clarity of communication in noisy or hectic situations. For example, Cassius Dio (early third century CE) wrote that the emperor Marcus Aurelius (161–80 CE), upon his return to Rome, addressed the crowd, which cried out the number of years he was absent from the city, eight, indicating it also with their hands, with the hope to receive this number of gold coins (*Epitome* 72.32). Placement of finger numbers in the public sphere indicates that they were common knowledge; both Pliny the Elder (*HN* 34.16.33; first century CE) and Macrobius (*Saturnalia* 1.9.10; fifth century CE) tell of sculptures of Janus indicating the 365 days of the year with finger symbols. Perhaps the most exciting evidence for the wide spread of finger numbers comes from circular *tesserae* of ivory or bone depicting hands with finger positions on one side and Latin, Greek, or both, numerals on the other (Figure 1.1). They present numerals from one to fifteen, as do similar *tesserae* with other pictorial representations (of persons, divinities, buildings, fruits, etc.), and most likely were game counters of an unknown game (Alföldi-Rosenbaum 1971).

Assigning specific words to numbers, thereby constructing some system of scale, is an important step towards a conscious appreciation of abstract numerical notions. While distinguishing small whole numbers up to five is an inherent ability of humans, assigning words to higher numbers or to parts (or fractions) indicates a culturally conditioned level of numeracy. As a general rule, the use of numerical terminology in documentary evidence, tomb inscriptions, theatre, hymns, poetry, or epic texts, may indicate levels of numerical knowledge in everyday life. In many Indo-European languages, the base for numeral words is decimal. In ancient Egypt, by the beginning of the third millennium BCE, individual Hieroglyphic signs were used to represent numbers for each power of ten (Figure 1.3; Imhausen 2016: 15–19). In ancient Greek, multiples of ten have distinct words up to the power of four (10^4), *myruas*, "ten-thousand"; and in Latin, up to the power of three (10^3), *mille*, "thousand." Apparently these magnitudes were common enough to create

FIGURE 1.1. Roman *Tessera*, finger counting, ivory. Bibliothèque nationale de France, Département Monnaies, médailles et antiques, Froehner.325.

specific words. In ancient Mesopotamia, by contrast, as early as the third millennium BCE, a sexagesimal place-value notation was used (Michel in Chapter 2 of this volume) and in cuneiform there are names for multiples of sixty to the power of two and three (60^2, 60^3), *šar*$_2$ and *šar*$_2$*-gal* respectively (Woods 2017: 428–9).

Fractions, more precisely parts, were common in ancient reckonings. In the second half of the third millennium BCE in Egypt, there were Hieroglyphic signs for "half," "third," "quarter," and "two-thirds" (Imhausen 2016: 52–4); yet, it is impossible to know how common these notions were among the general public. In the second millennium BCE, Assyrian merchants were customarily using parts to state prices in terms of silver, to convert values from one metal to another, and to calculate payments of interest rates on loans (Robson 2008a: 198–206; Michel in Chapter 2 of this volume). In the Greco-Roman world, calculating with parts was a widespread phenomenon. Generally, a unitary fraction system was used, in which all basic fractions had the numerator one (Maher and Makowski 2001: 378), with the exception of "two-thirds," which as early as the fifth century BCE had a distinct word in Greek, *dīmoiron* (Aeschylus, *Supplices* 1070). Latin, too, had a distinct word for "two-thirds," *bes*. For a short period *bes* probably signified a particular denomination, a small bronze coin minted in Rome *c.* 126 BCE by a certain Caius Cassius (RRC 266/3);[2] in the first century BCE, Cicero mentioned it to describe an increase in the interest rate from a third per month, i.e. a 4 percent yearly interest, to two-thirds per month, i.e. an 8 per cent yearly interest (*Ad Atticum* 4.15.7); and in the first century CE, it was used by Columella in his explanation of the division of the *jugerum*, the Roman standard land unit (*De Re Rustica* 5.1.11).

Calculational Tools and Artifacts

Numbers were reckoned and manipulated with fingers and hands also by using basic objects we generally call "counters" and their placement on counting devices of various sorts to accelerate calculations. Various calculating instruments were used daily in administration, commerce, dealings with authorities, organization of large-scale activities, teaching and education. For convenience, the Roman word ***abacus*** is used for all of these instruments, despite the fact that the abacus was hardly a Roman invention. Indeed, ancient abaci are attested cross-culturally and presumably predate the earliest textual and archaeological evidence, suggesting that their development was intertwined with the origins of various systems of numerical notation. They all "operate on the same basic principles of spatial reckoning in which numerical value is relative and positional within the device" (Woods 2017: 418–28, citation in 419); thereby, implying a level of abstraction.

There are two main types of abaci. The so-called "Eastern" abacus is a fixed-bead device made of wires stretched within or from a frame, along which beads are moved, either vertically or horizontally. Beads could be moved on the wires, but not from one wire to another, which results in a mental "carrying-over" of beads (or counters) from one wire (or column) to the other, making the operation of the device more sophisticated. The device is often portable. It is typical of East and Southeast Asia, but was also used in the ancient Mediterranean. From the Roman world, there are four known specimens of small bronze plates with slots that hold sliding beads and are marked with decimal place-value notation, which suggest that beads were moved vertically to the user (Figure 1.2; Schärlig 2001: 123–6). Though no abaci have survived from ancient Mesopotamia, drawing upon scant evidence in the textual sources and on cuneiform numerical notation, it has been suggested that early counting devices played a central role in Mesopotamian numerical life. A particular expression in cuneiform literary evidence, ***gu-dili-a***, which means "to string together on a single thread" or "to put in a row," is applied allegorically to tallying workers or counting people (Text 1.1), and suggests a counting device made of beads strung on hanging strings, probably inspired by the world of weaving and looming. Such a device, perhaps, is hinted by the paleography of the proto-cuneiform SANGA sign, which refers to the verb "to count" and later on to the profession of "accountant" or "expert scribe" (Woods 2017: 431–7).

The so-called "Western" abacus is made of a flat surface often marked with parallel lines, on which or between which counters are placed and moved. Movement is possible also between lines, based on definitional equivalences

FIGURE 1.2. Roman abacus, bronze, second century CE. Photo by SSPL / Getty Images.

between numbers according to the rules of replacement. Once the number of counters in a line or column reaches its maximum, it is replaced by one counter on the next, higher-order line or column. The counters themselves have no intrinsic numerical value, rather, value is inherently positional and relative. In counting boards of the ancient Mediterranean, lines usually were oriented vertically to the user and were based on a decimal place-value notation, often aided with an intermediated quinary lines or columns. One of the earliest examples is a fifth- or fourth-century BCE marble abacus from Salamis, which has Greek numerals on three sides of the board (Taub in Chapter 7 of this volume; Schärlig 2001: 61–71). Yet, numerals could be implicit rather than

nita munus-bi gu-dili-⸢a⸣ mi-ri-e$_3$-de$_3$ nig$_2$-u$_4$-da-ka-ne-ne-a ⸢igi⸣-zu-še$_3$ bi$_2$-ib$_2$-ta-la$_2$ [u]n šar$_2$-ra-zu kur$_7$-ak-da-bi dUtu-gin$_7$ igi-zu-še$_3$ i$_3$-dib-be$_2$-ne
You string together men and women on a single thread—their daily status hangs down before your eyes. Your numerous people parade before your eyes, as (before) the sun, for their inspection.

TEXT 1.1. *In-nin-ša$_3$-gur$_4$-ra* 211–212. After Woods (2017: 436, text no. 9).

written, with the advantage of using the device for calculating various scales of magnitude (fractions/parts, units, hundreds, or thousands) (Netz 2002: 325–9; Woods 2017: 418–20).

In the ancient Mediterranean, also non-specialists calculated by moving counters on an abacus or on an abacus-like surface. The practice was almost universal. Herodotus (2.36), when stressing Egyptian peculiarities, says that Greeks write and calculate with pebbles (*logizontai psefoisin*) from left to right, while Egyptian do so the other way around; calculating with pebbles was the trivial course of affairs. In late-fifth-century BCE Athens, Aristophanes makes his character Loathecleon ask his father to "calculate roughly, not with counters but on your fingers" (the phrase is as in Herodotus, *logisai . . . pseois*) (*Wasps* 656), suggesting that computations were typically performed with counters. In the Roman world, the abacus was in such regular use that it gave rise to a metaphor in one of Catullus' poems (first century BCE), where the poet counts the kisses of his lover as if on a "mental abacus," explicitly referring to errors of calculations or, perhaps, the shaking-up (*conturbo*) of an abacus, as an aid against those envious of their love (Text 1.2). Indeed, it is misleading to think of the abacus as a clearly defined artifact. Any pebbles may act as counters and any somewhat flat surface as a counting board; lines could simply be imagined. "The abacus is not an artifact: it is a state of mind" (Netz 2002: 327). Its power lies in representing numbers abstractly in spatial, non-linguistic terms. It allows breaking down calculations into a sequence of operations that require no effort, reducing it to an almost automated process of basic arithmetic (Netz 2002: 327; Woods 2017: 425–6). No ancient source discusses the abacus as such, probably because it simply was taken for granted. It was an omnipresent structure of thought.

Using abaci of whatever sort meant using counters in a specific setting of a counting device. Yet, in the ancient world, counters were employed in many settings, forms, and contexts; they were almost ubiquitous. In the Greek world, counters were applied customarily in the public sphere, facilitating all sorts of juridical and administrative procedures of the polis (Netz 2002; Cuomo 2012). And, in the Greco-Roman world at large, coins were the ultimate counters (Bailey 2012). To paraphrase Reviel Netz's comment, whereas we imagine numbers as an entity seen on the page, in antiquity they were imagined as an entity grasped between the thumb and the finger (Netz 2002: 329).

Writing Systems

The invention of numerical notation was created, almost certainly, along with the invention of script (Robson 2008a; Imhausen 2016: 16; Michel in Chapter 2 of this volume). Yet, writing systems that signify numerals often are

Vivamus, mea Lesbia, atque amemus, *rumoresque senum severiorum* *omnes unius aestimemus assis!* *soles occidere et redire possunt;* *nobis, cum semel occidit brevis lux,* *nox est perpetua una dormienda.* *da mi basia mille, deinde centum,* *dein mille altera, dein secunda centum,* *deinde usque altera mille, deinde centum;* *dein, cum milia multa fecerimus,* *conturbabimus illa, ne sciamus,* *aut ne quis malus invidere possit* *cum tantum sciat esse basiorum.*	Let us live, my Lesbia, and love, and value at one farthing all the talk of crabbed old men! Suns may set and rise again; for us, when the short light has once set, remains to be slept the sleep of one unbroken night. Give me a thousand kisses, then a hundred, then another thousand, then a second hundred, then yet another thousand, then a hundred. Then, when we have made up many thousands, we will confuse our counting, that we may not know the reckoning, nor any malicious person blight them with evil eye, when he knows that our kisses are so many.

TEXT 1.2. Catullus, Poem 5. English translation follows LOEB edition.

simpler than those used to record language, thereby allowing also less literate, or even illiterate, segments of society to understand written numerals. Some ancient systems that record numerals relied on processes of "subitization," i.e. the ability to perceive a small number of similar objects without counting it, and "chunking," i.e. the ability to organize large quantities into easily subitized groups (Woods 2017: 427–8). Basic knowledge of quinary and decimal systems eased the reading of numerals in several ancient writing systems: late Uruk unit fractions for grain (southern Mesopotamia, end of the fourth millennium BCE) (Robson 2008a: 292); early Egyptian records of spoils taken by Pharaoh Nārmer (*c.* 3150–3100 BCE) (Figure 1.3; Budge [1926] 1977: 5–6); later Egyptian mathematical texts such as the Rhind Papyrus (*c.* 1550 BCE) (e.g., problem 79; section titled "Writers of the Egyptian Mathematical Texts" in Michel in Chapter 2 of this volume); some cuneiform numerals, e.g. 𒐊 for "five" and 𒌋𒌋𒌋𒌋𒌋𒌋𒌋𒌋𒌋 for "ninety" (Woods 2017: 427); and Roman numerals, which formed a non-place-value system with a decimal and a quinary base, in which distinct symbols existed only for 1, 5, 10, 50, 100, 500, and 1000, while other numbers were written additively, e.g. XXXVIII for 38.

In ancient Greece and Rome, at start, the numerical symbols for numbers 1 to 4 were indicated by simple lines (I, II, III, and IIII), whereas some of the numbers greater than 4 were marked with the first letter of the number word (so-called "acrophonic" numbers). For example, in Greek: Γ (a graphic variant for Π) for *pente*, "five"; Δ for *deka*, "ten"; H for *hekaton*, "hundred"; and in Latin: S for *semis*, "half"; C for *centum*, "hundred"; M for *mille*, "thousand." In such a system, it was sufficient to know only several letters, rather than the whole alphabet, in order to read numbers. These systems imitate the operation of an abacus (e.g., ΔΔΔ|| for 32) and perhaps served to record results of calculations conducted tactilely on abaci or maintained abstractly on a "mental abacus." The Roman numeral system, in particular, emulates the results of an abacus operation, without needing to estimate clusters larger than five (Netz 2002: 327–9)

From the fourth century BCE, an alphabetic numeral system became frequently used in Greek documents. This was a decimal system, in which the Greek alphabet was divided into three sequences: *alpha* (α') to *theta* (θ')

FIGURE 1.3. Scene from the great mace-head of Nārmer, with a summary of the spoils taken by Nārmer in his wars appearing on the lowest row; after Budge ([1926] 1977: 5–6). Photo by Wikimedia Commons.

represent units from 1 to 9; *iota* (ι) to *koppa* (ϙ) represent tens from 10 to 90; and *rho* (ρ) to *omega* (ω) represent hundreds from 100 to 800, and, then, *sampi* (ϡ) represents 900 (Netz 2002: 328). With alphabetic numbers one can immediately recognize the order of magnitude according to the number of letters present (e.g., BKP stand for a number in the hundreds, namely 122). Since alphabetic numbers are independent of the spelling of the number names, they can be read even if one has only a basic knowledge of the alphabet and does not know how to spell the number names; just as one can pronounce the number 22 written in Arabic numerals without knowing how to spell "twenty-two." This makes numeracy possible even where literacy is low, especially when numbers are separately displayed in a text. Both acrophonic and alphabetic numbers, when inscribed on stone or bronze, were usually written in capital or "lapidary" letters; lower-case or "cursive" letters were reserved for the next level of teaching. Thus, learning the basic alphabet, without being proficient in literacy or receiving formal mathematical education, would suffice for "reading" numerals, as we learn from Hermeros, one of the dinner guests of Trimalchio, the notorious fictitious freedman in Petronius' *Satyricon* (Text 1.3).

Education

The spread of teaching numerals and numerical education in general are culturally and socially embedded. In ancient Mesopotamia and Egypt, mathematical education was mainly provided in schools intended for an elite group of initiates into the scribal profession, which, at least in Mesopotamia, could include both men and women (Robson 2008a: 97–106; Michel in Chapter 2 of this volume). In the Greco-Roman world, mathematical education was more widespread. Elementary education was conducted within the household under the supervision of the head of the family, who might acquire the service of teachers for the education of his children and slaves. Basic mathematical skills were an important part of elementary education of little

Non didici geometrias critica et alogias menias, sed lapidarias litteras scio, partes centum dico ad aes, ad pondus, ad nummum.	No, I never learned geometries, criticisms, and suchlike nonsense. But I know my block capitals, [and] I [can] say the one hundred parts of pounds, shillings and pence.

TEXT 1.3. Petronius, *Satyricon* 58. English translation inspired by Michael Heseltine's from 1913 in LOEB.

children in both Greece (Plato, *Rep.* 522c), where games were applied to teach numerical abilities (Plato, *Laws* 819a–c; Chapter 7 of this volume) and even slave boys could count and perform simple calculations (Plato, *Meno* 82–5); and Rome (Quintilian, *Institutio Oratoria* 1.10.35; Text 1.4).

Quintilian (first century CE) lists among the subjects in which boys should be trained before they are passed on to the *rhetor* for higher education music and geometry, both based on numerical reckoning (*Institutio Oratoria* 1.10.1–4). This followed Greek perceptions; however, Roman teaching of mathematics often took a less philosophical and a more practical direction. Cicero, for one, laments that Roman teachings of Greek geometry "have restricted this art to the practical purposes of measuring and reckoning" (*Tusculanae Disputationes* 1.5). The advantages from learning mathematics were not limited to obtaining applicable knowledge rather, the learning process of mathematical knowledge itself was appreciated as improving one's mind. Isocrates (fourth BCE) said that

Grais ingenium, Grais dedit ore rotundo *Musa loqui, praeter laudem nullius avaris.* *Romani pueri longis rationibus assem* *discunt in partis centum diducere. "dicat* *filius Albani: si de quincunce remota est* *uncia, quid superat? poteras dixisse." "triens." "eu!* *rem poteris servare tuam. redit uncia, quid fit?"* *"semis." an, haec animos aerugo et cura peculi.* *cum semel imbuerit, speramus carmina fingi* *posse linenda cedro et levi servanda cupresso?*	To the Greeks the Muse gave native wit, to the Greeks she gave speech in well-rounded phrase; they craved naught but glory. Our Romans, by many a long sum, learn in childhood to divide the *As* into a hundred parts. "Let the son of Albinus answer. If from five-twelfths one ounce be taken, what remains? You might have told me by now." "A third." "Good! you will be able to look after your means. An ounce is added; what's the result?" "A half." When once this canker, this lust of petty gain has stained the soul, can we hope for poems to be fashioned, worthy to be smeared with cedar-oil, and kept in polished cypress?

TEXT 1.4. Horace, *Ars Poetica*, 323–332. English translation follows LOEB edition.

by learning mathematics, young men exercise and sharpen their minds as "preparation for philosophy," but recommends that they study it only for a short term so that their minds will not be dried up (*Antidosis* 263–8). Four hundred years later, Quintilian echoed a somewhat similar view, that the "advantages [of learning geometry] come not as with other arts when it has been learned, but only during the learning process" (*Institutio Oratoria* 1.10.34).

Reciting mathematical lists was probably a commonplace of many ancient elementary education systems, evident in numerous written mathematical school exercises found around the Mediterranean (for Mesopotamia and Egypt, see Michel in Chapter 2 of this volume) written in Cuneiform, Hieratic, Demotic, Greek, and Latin, to mention only some of the leading scripts in use, inscribed on just about any writing material possible: clay tables, papyri, wax and wooden tables, and even limestone and bone. Yet, mathematical knowledge was achieved first and foremost by orally reciting multiples and divisions, as is evident in the humoristic verses of Horace (Text 1.4). Finger numbers, too, were a skill which took time to master; for example, in the time of Constantine the Great (fourth century CE), Firmicus Maternus wrote: "Do you see with what slow movements beginners in reckoning bend their fingers?" (*Mathesis* 1.4.13; English translation follows Williams and Williams 1995: 599).

KEEPING TIME

Timekeeping practices affect everyday numeracy in evasive manners, exemplified here by just two short remarks on civic aspects of timekeeping in the Roman world. On the construction of timekeeping artifacts, see Rochberg and Bergreen in Chapter 5 of this volume; on depicting these devices as representations of mathematical knowledge, see Taub in Chapter 7 of this volume.

Years, months, and days are cosmic time units, whose counting is governed by political and religious elites; but, they affect economic and civic life. In Rome, for example, since 304 BCE, a written version of the calendar was displayed in the Forum and inscriptions of Roman calendars, often called *fasti*, are preserved from the first century BCE onwards. In them, for each month, days of the *nundinae*—Rome's traditional eight-day week (nine days in Roman inclusive counting from one market day to another; Ovid, *Fasti* 1.54)—are marked by the letters A to H, signifying 1 to 8. The *trinundium*, a period of three *nundinae*, was a regular interval for notice-giving in Republican politics, in legislative procedure, and in Roman law and business. And, the *nundinae* framed even private routines (Ker 2010: 364–70), such as cutting one's nails (Pliny, *NH* 28.28), bathing (Seneca, *Ep*. 86.12), and shaving (Varro *Sat. Men.* 186, *apud* Nonius 316 L). Next to each day, further letters mark the Kalends (KAL), the first day of the month; the Ides (IED), the fifteenth or thirteenth day of the month; and the Nones (NON), eight days before the Ides (nine in Roman inclusive counting). In the literary and epigraphic record, Romans often indicated dates by counting the number of days before or after one of these three time-indicators. Also marked in the public calendars, is the Roman division of days into *fasti* (F), days in which private and public business could be conducted, including the operation of law courts; *nefasti* (N), days in which business could not be conducted and religious rituals were ordained; and comitiales (C) days in which a motion could be brought before the people (Varro, *Ling.* 6.30; Macrobius, *Saturnalia* 1.16.2–5, 14–15; Hannah 2009: 27–8). Such an abbreviated display allowed also those who only "knew their lapidary letters" (Text 1.3) to follow the calendar, especially where inscriptions contained a place for a moveable peg, which, when moved daily, helped also the less literate to follow the counting of days and plan their affairs accordingly.

In the Greco-Roman world, many people could count their age and age is frequently referred to on tombstones and in documents for identification purposes. A tendency for (a quinary) age-heaping is well attested in the epigraphic record and analyzed statistically to estimate levels of numeracy (Duncan-Jones 1977; Scheidel 1996; Baten and Priwitzer 2015). Nonetheless, precision in stating one's age is also attestable; for example, a Roman lawyer from Dalmatia lived 47 years, 9 months, and 7 days (*CIL* 3.2127a). Others were not so well informed but, nevertheless, had a desire for numerical precision in stating an exact age; for example, Otacilia Narsica dedicated a tombstone to her husband who lived "more or less 28 years" (*CIL* 6.20307: P-M-ANNIS/XXVIII). The papyrological record suggests that the context in which age was stated affected punctuality: where age is used for general description, such as in contracts, age-heaping is commoner; where age mattered financially or could be checked against previous declarations, such as in census declarations, it tended to be precise (Scheidel 1996: ch. 2).

MEASURING VALUE

Measuring value is crucial for social and economic conduct such as, inheritance, commerce, and domestic accounting. Understanding quantities and possessing numerical skills dictate the modalities of economic interaction and affect the evaluation of wealth. Metrological systems tend to expand from one sphere of life to another and the numerical magnitudes which they use are applied in different contexts.[3] Units for measuring weight become units for assessing value and may act as monetary units of account; for example, the Mesopotamian *shekel*, the Greek *obol*, and the Roman *uncia*. Ancient metrological and monetary systems used a variety of scaling methods, mainly decimal and duodecimal, but also a mixture of both reckoning systems together with a hexadecimal one. Here, focus is on the use of numerical terms by the less educated inhabitants of the Greco-Roman world, and how easily did they conduct calculations and at what level of sophistication.

The invention of coinage at the end of the seventh century BCE in Lydia opened a new era in the relations between measuring value and applied numeracy. Coins were a new media that tangibly communicated to a much wider audience numerical reckoning in the measuring of value. Coins fit nicely with Greek "counter culture" (Netz 2002: 329–34) and spread rapidly in the Greek world: three generations after the first coins were struck, dozens of *poleis* were minting coins. More often than not, money was now being counted rather than weighed. Neither special skills of alloy assessment nor designated tools such as scales were required for using coins in exchange; it sufficed to know basic counting (Schaps 2004: 111–23). In Aristophanes' comedies (fifth century BCE), coins commonly appear in numerical evaluation of commodities (*Ach.* 960–2; *Frogs* 549–54) and services (*Wasps* 1125–8). Jokes regarding the cost of the moonlight (*The Clouds* 607–14) or that of delivery services to the underworld (*Frogs* 170–8) indicate that the relative value of things expressed in numerical monetary terms was common knowledge.

Surely, not all transactions were carried out numerically and not everyone was accustomed to such reckoning. Pericles' (*c.* 495–429 BCE) habit of selling all the produce of his estate in the *agora* and run his household on the basis of daily monetary transactions was the exception rather than the rule for elite Athenian housekeeping (Plutarch, *Per.* 16.3–5). At the beginning of the fourth century BCE, the Pythagorean Archytas of Tarentum extended the benefits achieved from calculation (*logismos*) in private transactions to the running of the citizen body, stating that they applied to both those who knew and those who did not know how to calculate (DK 47b3[4]; Seaford 2004: 269). Yet, from the numismatic and literary records, it is clear that coins were regularly used by many people from different strata of Greek society. The frequency of using coins as tangible counters with which value was calculated, enabled comparing the value of things that belonged to different spheres of life on an equalizing numerical scale and had deeply affected ancient Greek culture (Kurke 1999b; Schaps 2004; Seaford 2004).

In the two generations that followed Alexander the Great's conquests, the monetization of enormous stocks of precious metal stored in the coffers of Egypt and the Near East resulted in a wide distribution of currency. The role of coinage as a means for counting and assessing value had introduced Greek numeracy as a new mentality in those areas where the use of coins was advanced as a vital instrument of Hellenistic governance and taxation (von Reden 2007). Rome's taking over of the Hellenistic world did not abolish local currencies, rather, it gradually turned the Mediterranean into a jigsaw of compatible monetary systems. Numismatic evidence displays the variety and complexity of archaic and classical Greek denominations, and later those of the Hellenistic and Roman world. Some of the names for denominations reveal their worth by using numerical concepts, such as multiples and parts,[5] as do the relations between the coins' diameter, which helped to distinguish between denominations. Tokens, first introduced in the fifth century BCE, brought a new level of abstraction to the use of denominations as representatives of numerical value; as did the use of bi-metallic and tri-metallic currency systems.

Monetary denominations ultimately rely on abstract mathematical ideas, ratios, and multiples. However, possessing situated and procedural knowledge of how to use money, is not identical to holding a conceptual knowledge of the abstract meaning of numbers. In the Greco-Roman world, everyday use of money in commercial activity required subtraction and addition of whole numbers and of parts. The use of unit fractions that characterized Greek denominations, also dominated the Roman monetary environment and Roman calculation methods. Calculational routines usually were not conducted in writing; they were carried out mechanically on abaci or abacus-like surfaces using coins as counters and frequently relied on division tables and their oral memorization (Text 1.4; Maher and Makowski 2001: 390–3).

Melissa Bailey (2013) divided monetary numerical practice into two categories: transactions carried out with actual coins, and those recorded with written monetary units of account. Focusing on the Roman world, she combined numismatic, material, literary, and documentary evidence (graffiti and written records of financiers) from Pompeii and demonstrated that, in physical daily encounters, monetary amounts were expressed with those specific coins most common in the material record. In such transactions, the numerical habit was often one of singular addition and subtraction, based on low, round, or repetitive numbers, with an overall emphasis on simplicity. Bargaining probably was based on loose notions of the proper amount that corresponded to a given number of coins in a specific denomination. It seems that for people with little formal training, coined money defined number and numeracy. This is supported by price evidence for wheat from Roman Egypt, which tends to be limited to multiples of the *tetradrachma* (e.g., 12, 16, 20, 24 *drachmas*), the main denomination of the province (Rathbone 1997).

The other monetary numerical practice regarded money in written units of account, usually expressed in *sestertii*, a quarter of a *denarius*. By the first century BCE, Roman monetary system was mainly a quaternary system: four *quadrantes* to the *as*, four *asses* to the *sestertius*, four *sestertii* to the *denarius*; as well as a decimal one: twenty-five *denarii*, which equal a hundred *sestertii*, to the *aureus*; the maximum legal interest rate being the *centesima*, the hundredth part per month, which meant a 12 per cent interest rate per annum. Yet, in accounting and in calculating interest, Romans resorted to the traditional duodecimal system. Especially in contexts of loans, inheritances, and property, money was conceived as a mass that can be divided according to the standard duodecimal fraction division (Cicero, *Ad Atticum* 7.8.3; Justinian, *Institutiones* 2.15.5); and its manipulation required more specialized methods of calculation, that involved the abacus, finger numbers, and fractional tables, either written or memorized. Not everyone was accustomed to this type of numerical practice and the less educated were proud of possessing such a skill; for example, the freedman Hermeros (Text 1.3), or a tombstone of a twenty-year-old man who is praised for being able to calculate up to the 300th part (*ILS* 7753), that is, a third of a one-hundred, which in Roman reckoning amounts to a 4 percent yearly interest.

PRACTICING POLITICS

Kingdoms and Taxes

Dealing with the ruling authorities involved paying taxes and rents. Whether in monetary form or in kind, these were assessed as quantities or sums and calculated in some numerical way, which in the ancient world often was termed as parts, for example, a tenth of the yield on arable land, an eighth from the orchard, etc. In ancient societies, almost everyone had direct or indirect daily, monthly, or yearly dealings with numerate professionals such as tax collectors and landlords. To look after their economic interests, people needed a minimum of numerical knowledge, or to have trust-based relationship with a person who had such knowledge. Land surveying, essential for working out taxes, rents, and other economic affairs, required professional knowledge. Still, an ability to appreciate the meaning of land assessment and perhaps a basic capacity to roughly measure landed property was acquired by landowners, and in the Greco-Roman world, also by lessees and tax payers. Tenancy contracts from Roman Egypt show local landowners and tenants habitually using numerical terms (Rathbone 1991; Kehoe 1992).

When central authorities impose communal taxation, local authorities negotiate tax collection and social formation within the community lessens the need for individuals to possess better numerical knowledge. When responsibility for paying tax is delegated to individuals, their numerical abilities are likely to accelerate. The spread of monetization during the Hellenistic age brought with it new forms of taxation. The best recorded example is Ptolemaic Egypt, where new coin-based taxation was combined with a millennia-long history of central control on populace and land. From the third century BCE, a monetary poll tax was imposed in the form of a salt tax (*halikē*) payable by almost all inhabitants of Egypt, attested during the third century BCE for both males and females at decreasing rates (Clarysse and Thompson 1995). Surely, some people paid their poll tax via labor and patronage relations. Still, reckoning monetary tax became a commonplace for wider segments of society: it was included in ordinary marriage contracts as part of the husband's obligations to his wife (BGU 5.1463 (Elephantine, 246 BCE)); and was part of routine arrangements between landlords and tenants, reckoned for in accounting books at the end of the rental year (von Reden 2007: 84–110). Outside Egypt, information on individual rather than communal taxation is relatively scarce. However, sporadic references to "head tax" (*epikephalaion*) and occasional exemptions from it, such as royal gifts of tax-free movements (e.g., *1 Macc.* 10.29, 10.34), indicate that just about everyone was exposed to numerically based taxation (Kaye 2018).

In addition to rents and taxes, empires may influence numeracy in all sorts of ways. Rome, for example, forcibly extracted her imperial power over her former enemies, initially in the Apennine Peninsula and later in other regions of the Mediterranean, by disseminating Roman culture in architecture, commercial institutions, laws, and the Latin alphabet and language. Numeracy, too, played a role in relations between Rome and her allies and subjects. In a probably fictionist debate between Augustus' two senior advisors, Agrippa and Maecenas, allegedly taking place in 29 BCE and written in the early third century CE, Maecenas advises young Octavian (later to become Augustus) that, "None of the cities should be allowed to have its own separate coinage or system of weights and measures; they should all be required to use ours" (Dio 52.30.9). Although Augustus did not enforce the use of the Roman measurements and monetary system as a compulsory measure for all of the empire's inhabitants, by the time Dio was writing, reality was very much attuned with Maecenas' advice. The processes by which such a policy was put to work was often gradual, spanning over decades, sometimes even centuries; in Roman Egypt, the local monetary system was abolished only in the time of Diocletian (296 CE). At other times, Roman policies for advancing its cultural hegemony were more fiercely enforced. For example, in the *mensa ponderaria* at Pompeii, a counter used to check capacities for liquid and solid foodstuff, Latin inscriptions appear next to the measuring bowls. However, traces of three Oscan inscriptions indicate that both measurement cavities and text were reformed to Latin standards, probably in the aftermath of the Social War after the city was refounded as a Roman colony in 89 BCE (Figure 1.4; Cuomo 2012: 4–5). It appears that units of measurement, too, did not escape "Romanization."

Citizens and Voting

In the ancient Mediterranean world of city states, the registration of citizenship had practical implications on everyday life of communities and individuals. Phoenician, Greek, and Italian citizenship was accompanied by a set of rights and obligations, which more often than not were related to wealth assessment. As such, citizenship was recorded and supervised by both state authorities and the citizens themselves. In political systems where larger segments of the citizen

FIGURE 1.4. *Mensa ponderaria* (measuring table) from Pompeii, first century BCE. Photo by The Forum / Flickr / Wikimedia Commons.

body were involved in public affairs via elections and assemblies, voting procedures that included counting were part of the discourse in the public sphere. In Greek democracies, Athens being the best documented exemplum, and in republican Rome, and in other less documented Italian communities, voting assemblies were almost a daily feature of public life. Democracy, whether direct (e.g., in Athens) or representative (e.g., in republican Rome), brought with it the intensification of applied numeracy among non-elite members of society who participated in public politics.

The method by which voting units were divided and the criteria for allocating citizens to voting units were repeatedly negotiated in city-state societies of the first millennium BCE. Rome provides two well-known examples: first, in the early Republic (6–4 BCE) developments in the complex formation of the timocratic *comitia centuriata*, which was comprised eventually of 193 voting units, were intertwined with contemporaneous social struggles between Patricians and Plebeians (Cornell 1995). Second, in the aftermath of the Social War (91–89 BCE), after the massive grant of Roman citizenship to Italian allies, fierce political struggles took place regarding how to register the new citizens in

Rome's 35 tribes, which affected their potential influence in the *comitia tributa*, with the situation resolved only in 70 BCE (Velleius Paterculus 2.20; Appian, *Bellum Civile* 1.49).

Counting the votes, too, was a public affair. In Athenian democracy, for example, there were two main forms of vote counting: usually the assembly would vote by show of hands, with ostracism applied in exceptional circumstances; while in law courts and on other occasions, votes would be carried out by counters (Netz 2002: 336; Cuomo 2013). In fourth-century BCE Greece, enemies of democracy showed their contempt for it by calling it a system of governance subject to "arithmetical proportion," as opposed to an aristocratic system subject to a "geometric proportion," by which they meant that higher representation would go to those abler to support the polity (Plutarch, *Moralia*, 719b–c = *Quaest. Conviv.* 8.2.2). Although this metaphor is a legacy of a smirking, mathematically educated elite, it brings out an important fact, namely, that in political systems where larger segment of the citizen body was actively involved, literally, "every vote counted."

Managing the public budget sometimes was also negotiated by commoners. In Athenian democracy, financial accountability of public magistrates was the *bon ton* of the day. State officials who completed their term had to render accounts publicly, on the basis of which they could be challenged by any citizen (Cuomo 2013). The scribal habit of displaying numbers in fifth- and fourth-century BCE Athenian inscriptions is evidence of the importance of basic numeracy in Athenian democracy. While the default setting for displaying numbers seems to be an interspersed format, which is less focused on conveying the numerical information, in inscriptions dated to the height of Athenian democracy, between 460 and 420 BCE, numbers are often displayed in a more easily read tabular manner, which makes numerical information accessible also to the less literate and fits well with the Athenian democracy's inbuilt mechanisms of accountability (Cuomo 2012: 10–11; 2013). That calculating such high public finance was accessible to the general public is attested to in one of Aristophanes' comedies (*Wasps* 655–722). Even if these verses were a parody on contemporary rhetoric, in which orators integrated in their speeches arithmetical calculations of various accuracy, not always fully followed by their audience, the rhetorical use of calculations generally was in accord with democratic values and often appealed to commonsense and everyday experience, enabling citizens to keep track of it (van Berkel 2017).

Lastly, the battlefield, too, required numerical knowledge. In classical Greece, generals were expected to reckon their campaign expenditure (*Inscr. Graec.* I^3 71, lines 45–50 (425/4 BCE); Cuomo 2013: 260). Plato says plainly that there is need to "set down as a study requisite for a soldier the ability to reckon and number (the verb is *arithmeīn*) . . . if he is to know anything whatever of the ordering of his troops, or rather if he is to be a man at all" (*Rep.* 522). Military commanders responsible for strategic planning and troop advancement and placement were expected to have a numerical education. In the second century BCE, Polybius (9.14) wrote that because days and nights are of unequal length both astronomy and geography are needed to calculate the distance traveled in a day's or a night's march. Some level of numeracy was required also in lower ranks of command, down to the regular soldier. Mercenaries and soldiers who received monetary pay were able to follow their salary. In the Roman army, lower ranks of command were often literate and could handle basic calculations (Bowman 1994, for Roman Britain; Baten and Priwitzer 2015, for Pannonia) and numbers were habitually used in Roman military communication lines (Remijsen 2007). Good knowledge of geometry was required in building Roman military camps, which were of a uniform scheme, a fact that helped soldiers to find the chain of command and their corps within the camp (Polybius 6.27–32; 6.42; Cuomo 2001: 64–6). In the late fourth century CE, Vegetius credited Rome's military supremacy to better training and education of its soldiers, stating that the administration of a legion required soldiers who were acquainted with *calculandi computandique* (reckoning and calculating) (*Epitoma rei militaris* 2.19).

HAVING FUN

Pleasure and leisure activities are integrated into everyday life, just as practical and useful activities are. As their aim is to produce enjoyment and delight, the extent to which games contain numbers, numerals, estimates of sizes or probabilities, and other forms of calculation, is good evidence for levels of numeracy. Because games are intended to be fun, usually the mathematical knowledge they assume participants to hold is such that could be applied without much effort. People play games because they take pleasure in exercising skills in which they are already somewhat proficient.

Board and dice games are leisure activities that rely on the understanding of numerical concepts and often require an ability to conduct basic calculations. Such games are evident in archaeological and literal records from Egyptian, Mesopotamian, Greek, and Roman antiquity and were enjoyed by all segments of society. For some games, we have only mute archaeological evidence, which tells little about how they were played, for example Roman circular *tesserae* depicting numerals from 1 to 15 on one side and different pictorial representations on the other (Figure 1.1). For other games, the archaeological record is supplemented by written evidence, which allows partial or full reconstruction of the game in question. The close relationship between board games, mathematics, and accounting systems has long been observed. The widespread use of board games and their "lower" social status, meaning that they were played by just about everyone, makes them helpful for estimating levels of numeracy of the non-elite.

The first board games were race and alignment games and were played in Egypt, the Near East, and Mesopotamia. The most famous and widespread of these are the Egyptian game of Senet and the Game of Twenty Squares, which in its earliest, third-millennium BCE version is known as the Royal Game of Ur. The latter was played on a board consisting of a larger "body" of 3×4 squares, then a "bridge" of two squares and a smaller "body" of 3×2 squares (Becker 2007: 11). Early in the second millennium BCE, the board was changed to a "body" of 3×2 squares with a long "tail" of 8 squares (Figure 1.6); and this version became common in the second and first millennia BCE throughout Mesopotamia, Egypt, and the Near East, with specimens found also in Asia Minor, Cyprus, and Crete (Voogt, Dunn-Vaturi, and Eerkens 2013). A late version of the rules of this game is preserved on a Hellenistic cuneiform ostracon (BM 33333B; Finkel 2007a). The Egyptian game of Senet was played on a board scored as lines forming a 3×10 grid of squares, on which draughtsmen were moved horizontally in the same direction on each row until the last square was reached. The Senet board was established as early as the first dynasty (*c.* 3100–2900 BCE), the earliest full description of the game was found in the tomb of Hesy-Ra from the third dynasty (*c.* 2670–2600 BCE), and the latest Senet board found in Egypt is dated to the Roman period (first century CE) (Piccione 2007: 54–6).

Both Senet and the Game of Twenty Squares were race games played by two players on a board using draftsmen and a randomizing agent of some sort, where competition was based on both strategy and luck. Both games were continuously played from the third millennium BCE until the Roman period (see Figures 1.5 and 1.6). Both were associated with divine and symbolic meanings (20 being the number of the Sun God Shamash and is otherwise related to Mesopotamian divination; Becker 2007; Senet was associated with the migration of the soul in the afterlife and with funerary rites; Piccione 2007: 58–61). And, both were continuously played at all levels of society, as attested by well-carved game boxes deposited in tombs as well as by unpolished portable specimens crudely carved on cheap material found sporadically in settlements. The widespread and continuous dispersion of these board games, as well as other games not specified here, indicate that computations up to thirty, combined with assessments regarding the possibilities offered by a randomizing numerical element and estimates regarding the opponent's movements, were easily and enjoyably performed by game players from all strata of society.

In the Greco-Roman world, board games played with counters, with or without the use of a randomizing agent, were extremely common. The assumption that we can always determine which ancient artifacts acted as calculative boards (abaci of some sort) and which as game boards, is misleading. There is no straightforward boundary separating artifacts into distinctive usages and this observation applies also to counters (Greek *pessoi*, Latin *calculi*) (Netz

FIGURE 1.5. Bifacial board game of the Game of Twenty Squares (on its reverse there is a board of Senet in its 3×11 version). Hazor, Israel; approx. ninth century BCE. Photo by Mariana Saltzberger, courtesy of the Israel Antiquities Authority.

FIGURE 1.6. Bifacial board game box of Senet and the Game of Twenty Squares. New Kingdom, 1550–1295 BCE. Photo courtesy of the Gift of Egypt Exploration Fund, 1901, The Metropolitan Museum of Art.

2002: 327; Purcell 2007: 91–2; Cuomo 2012: 4). Greco-Roman dice and board games were played at all levels of society, either on private boards or boards inscribed on immovable surfaces in the public sphere. The latter became an integral component of the Greco-Roman urban environment and are found throughout the Mediterranean, either as designated tables or, more frequently, inscribed on pavements, floors, ceilings, and stairs of public buildings and temples (Bell and Roueché 2007). Yet, we know little about how most of them were played. A popular Greek game called *polis* was a strategic game played without dice on a grid of squares of unknown size with each of the two players having many pieces, perhaps 30 (Pollux, *Onomasticon* 9.98). Somewhat similar was the Roman game *ludus latrunculorum* ("the Game of Brigands"), played by two players with many counters (called *latrones*, *calculi*, *bellatores*, or *milites*) and without a dice on a grid of squares of varying size (common specimens have 8×8, 10×10, or 15×15 squares), depending on the number of draughtsmen (*Laus Pisonis* 192–208). Common race games include the Greek game *pente grammai*, "five lines," played by two players with dice and 5 counters per player on a board made of five lines (Kurke 1999a); and the Roman game of *xii scripta*, "the Game of Twelve inscriptions," played by two players with dice and 15 counters per player on a board of 3×12 placements separated in its middle, with placements indicated by circles or letters, which sometimes even form a hexameter (Ovid, *Ars Amatoria* 3.363). All of these games show that in the Greco-Roman world, like in the Egypt and Mesopotamia, counting and performing basic calculations within the realm of 30 was a common and well-mastered ability which enabled people from all strata of society to participate in these playful activities.

Accompanying board games, especially those involving a randomizing mechanism, is betting on the results. Gambling was a leisure activity enjoyed by commoners as well as members of the elite (Tait 2007: 46–51, for Egypt; Suetonius, *Augustus* 71; *Claudius* 33, for Rome). The numerical abilities required for dice games and gambling need not be rationalized fully. For game players in antiquity, chance was not a purely mathematical property, rather, a randomizing factor (dice, throwing sticks, astragali, etc.) was connected with divination. While dice games do not require profound mathematical knowledge, a basic recognition of numbers is indispensable to play them. Whereas dice and board games were played by just about everyone, mathematical riddles, often in the form of poems, were probably enjoyed mainly by an educated elite. Some Greek mathematical epigrams show highly complex problems; others give relatively simple riddles that tell of counting bowls or fruits (*Greek Anthology* 14.118) and fit with Greek methods for teaching children (Plato, *Laws* 819a–c); there are even burial epigrams (*Greek Anthology* 14.125–6), perhaps publicly posted on tombstones. All of these riddles required calculating a sum on account of knowing proportions of its parts (Taub in Chapter 7 of this volume).

CONCLUSION

In antiquity, numerical terminology based on a decimal system was well known and widely used, as was terminology for basic unitary fractions, perceived as parts ("half," "third," "quarter," "sixth," "tenth," with "two-thirds" being the exception). Basic calculational tools, i.e. counters, pebbles, and different sorts of abaci, were used from the dawn of civilization. In Egypt and Mesopotamia, different elite groups habitually used these artifacts, whereas in the Greco-Roman world, calculational artifacts were used by all segments of society, sometimes on a daily basis. Concerning mathematical education, in Egypt and in Mesopotamia this was mainly the privilege of elite groups and sometimes also of "lower elite," such as Assyrian merchants in the mid-second millennium BCE. In the Greco-Roman world, counting and performing basic calculations within the realm of a hundred was an integral part of the basic education of children from just about all segments of society.

As to the questions of who was numerate and why: for Egypt, we have very little evidence to indicate levels of numeracy beyond the educated elite. The case for Mesopotamia is only somewhat a little more revealing, with evidence of Assyrian merchants in the second millennium BCE and some domestic accounts in the first millennium BCE shedding little light on the numerical abilities of people who belonged to a social standing not as high as the elite or scribal professionals, but higher than the average illiterate person. In the Greco-Roman world, with its civic mentality and more available evidence, we are in a better position to inquire into levels of numeracy of the general public. From the third century BCE, yearly calendars were publicly known and dictated which days were forbidden for civic management and economic conduct. Surely in urban environments, a social time frame affected weekly customs and daily routines also of illiterate segments of society and was implemented in their basic calculations (adding and subtracting) within the realm of a dozen (the hours of the day), and even a hundred (counting one's age). The spread of coinage in the Greek world and the almost omnipresence of coins in the Hellenistic and Roman worlds brought applied numeracy to the forefront, with coins defining numbers for those with little numerical education. For those who were somewhat better educated, including not only members of elite groups, calculating with parts, using a duodecimal division system, and counting within the realm of the hundred, was probably widespread knowledge. Imposing individual taxation by central or imperial authorities and participating in citizen-based political systems, in which voting was an important part of public routines, enhanced the use of numerical terminology and of reckoning to wider segments of society. Lastly, the widespread and perpetual distribution of boards games indicates that counting and performing basic calculation within the realm of thirty was common knowledge in all segments of society in Egypt, Mesopotamia, and the Greco-Roman world.

CHAPTER TWO

Practice and Profession

Everyday Mathematics and Their Transmission in Mesopotamia and Egypt

CÉCILE MICHEL

The mathematics we use today has its roots in antiquity and, in various aspects, more specifically in Near Eastern antiquity. Their inventors and practitioners have left texts—only a small portion of which have survived to date—that reflect their knowledge and the practice they made of mathematics. However, these texts are unevenly distributed in time and space, partly because of their more or less durable medium.

In Mesopotamia and the surrounding areas, more than a million cuneiform texts, the great majority written on clay tablets, but also on stones or other media, have been discovered. They cover more than three millennia from the end of the fourth millennium BCE to the first century CE and were found in a very large area from the Mediterranean Sea to the Persian Gulf and Iran, from Anatolia to Egypt (El Amarna).[1] Some sites have produced only a few tablets, while others tens of thousands of them. About a dozen languages have used the cuneiform script, the main ones being Sumerian and Akkadian. Assyriologists consider that about 2,000 of these texts have a mathematical content, most of them were found in an educational context. But tens of thousands of other clay tablets linked to the administration of palaces and temples or to private economic activities contain quantifications and computations and thus may be linked to the practice of mathematics.

During roughly the same period in Egypt, besides monumental hieroglyphic inscriptions engraved or painted on the walls, surviving texts are written in

Hieratic and Demotic, mainly on papyri and ostraca. Ostraca are numerous from the middle of the second millennium onwards; there are almost 10,000 ostraca written in Demotic. Papyri hardly survived time compared to clay tablets, and we count about as many Egyptian Hieratic papyri as cuneiform mathematical texts. For the first millennium, there are a little less than 5,000 papyri written in Demotic.[2] Taking apart the ostraca, for which there is no evaluation of the number of texts with a mathematical content, the corpus of Egyptian mathematical texts represents about 1 per cent of the Mesopotamian one. This disproportion in the available sources is reflected in this chapter, which focuses on Mesopotamia with some hints of mathematical practices in Egypt. Such a small number of mathematical texts might be explained by the fact that the context of discovery of the Egyptian papyri is limited to graves and temples.[3]

Most of the authors of these mathematical texts remain unknown. Were they professionals of mathematics? A "profession" is understood as a paid activity performed by a person who went through a specific training or education. There is no word in Sumerian, Akkadian, or Egyptian that we would translate as "mathematician." In contrast, there are words to refer to "astronomers." In Akkadian, astronomy practitioners are called *ṭupšar Enuma Anu Enlil*, "the scribes of the Enuma Anu Enlil" which is the longest Babylonian treaty of astrology, written on 70 tablets. In Egyptian, *imy-wnw.t* is "the one in charge of the hours," generally translated as the priest-astronomer. The absence of a word for "mathematician" reflects the absence of mathematical professionals despite the presence of mathematical problems whose resolution required specific and advanced training. Mathematics in antiquity was studied and subsequently taught and developed by masters. Those masters, however, were equally knowledgeable in various other "sciences" as astronomy, medicine or divination as we gather from their archives or libraries.

Besides "mathematical texts," thousands of administrative and economic texts from various cultural environments testify to the use of mathematics outside the scholarly milieu. Quantification and computation are practiced daily by many actors, whether they are employed by institutions, palaces or temples, or act as independent workers, such as merchants and some craftspeople. Since our main sources of information concerning these practitioners of mathematics are textual, their identification is unavoidably linked to their ability to write.

For more than a century, it was generally accepted that literate and numerate people in the ancient Near East represented a small elite. However, for two decades, in-depth studies on tablets found in the context of "schools" at Nippur dating to the beginning of the second millennium, on the materiality of texts and on paleographical analyses of cuneiform tablets, have highlighted several levels in the practice of writing. School masters or scribes working for the powerful have been through the complete scribal curriculum reconstructed by

Assyriologists. They were the authors of advanced mathematical texts, of official letters and treaties, literary composition, astrological and divinatory treaties. A second group of people only finished the elementary level and could write family and commercial contracts as well as private letters; they were able to solve simple mathematical exercises. Others learned to read, write, and count at home from a family member.

Depending on the periods, regions and milieus considered, the cuneiform syllabary used to write Akkadian had different degrees of complexity, varying from 100 to more than 250 signs, including some Sumerian logograms. Moreover, the technique of cuneiform writing is very simple: each sign for logograms, syllables, and numbers consists of a composition of wedges, each wedge being obtained by impressing a stylus in fresh clay. Numbers are written with only two types of wedges (vertical and chevron). For all of these reasons, literacy and numeracy, at least during the first half of the second millennium, was quite widespread among both men and women. During the first millennium, the Aramaic language was gradually gaining ground. It is written in ink on papyrus or other perishable media. The Akkadian language written with cuneiform signs on clay tablets remained the scholarly language in cultural and religious contexts. Because of the lack of texts of practice, it is difficult to evaluate the proportion of the literate and numerate population.

In Egypt, Hieroglyphs were reserved for monumental writing. During the third and second millennia, hieratic was the cursive everyday script, and it was usually written on papyri. Since this media hardly survived time, there are too few texts to answer the question of literacy and numeracy. Thanks to the use of demotic written on ostraca and limestone during the first millennium, many more texts have been found. However, there is still no clear idea about the number of literate and numerate Egyptians for this period.

Despite the abundance and variety of sources available, at least for ancient Mesopotamia, the texts document the type of mathematics involved, but rarely the authors of computations. Most often, these texts not being signed, the identity of mathematical practitioners must be guessed at only through their activities and the context in which the texts were discovered (Michel and Chemla 2020a). It is then possible to consider their role and status in society. Naturally, only the individuals who left behind archives are documented, which induces a bias in our study. This is especially true for third- and second-millennia Egypt, where only members of the upper class have benefited from decorated tombs, in which, occasionally, some papyri were discovered.

The practitioners of mathematics in the ancient Near East came from a great variety of milieus and the mathematical methods they learned and used varied accordingly. These methods were taught in "schools," in private context by masters, or passed on from generation to generation. Mathematics usually played an important role in the education.

MATHEMATICS IN PRACTICE AND THEIR PRACTITIONERS

Many fields of daily life activities in Antiquity required some knowledge in mathematics. Their actors, whether officials working for palaces or temples, or independent workers, merchants, and craftspeople, could be numerate enough to proceed themselves to the mathematical operations necessary for their activities. Such a study implies gathering dossiers, outside the corpus of mathematical texts, including administrative and economic texts, as well as letters. It is often difficult to draw the portrait of these mathematics practitioners and to distinguish between scribes or high-level officials with multiple competences and specialists under their authority. Moreover, many of these individuals remain anonymous. However, this section is an attempt to highlight the role and status of some of them in society. Leaving aside the writers of mathematical texts who are at the center of the next section, a few samples of practitioners of mathematics borrowed both from Mesopotamia and Egypt are presented here; they deal with land surveying, the management of estates, trade, and architecture.

Mesopotamian Land Surveyors

Agriculture was the main resource of ancient Mesopotamia. The administration of land was centered on the temple or royal estate. Land measurement with rope and rods of different lengths is well developed during the third millennium. In the south of Mesopotamia, the necessary irrigated agriculture had an impact on the shape of the elongated fields, their smaller side receiving the water from the irrigation channel. Beside a heavy infrastructure for the building and management of dams, dykes, and canals reflected in mathematical texts, specific personnel were dedicated to the measurement of fields and the estimation of the cultivating area. The archives found at Girsu around 2350, and belonging to the domain administered by the queen, illustrate the land surveyors at work (Lecompte 2020). To measure fields, land surveyors did not take into account their shape, but proceeded with length-measurement values and approximations. They used the "land surveyor formulae," i.e. the product of the average of the irregular quadrilateral sides. Camille Lecompte has identified three different professions linked to land surveying and registration: the land recorder, the geometer who proceeds to the measures, perhaps together with the field assessor. In this hierarchical organization, it is most probable that the land recorder was the one to proceed to the computations.

At the end of the third millennium, in order to face the management of new territories, the Ur III state implemented extensive administrative and economic reforms. An increased number of scribes worked for the state, recording daily accounts, then summarized in annual balance accounts based on estimated theoretical expenditures. Among the recorded data was a precise quantification of irrigated arable land. Stephanie Rost presents the case of a large domain, Da-Ummaki, located in the Umma province which was under the responsibility of two families of land surveyors, in which skills were transmitted from father to son (2020). Descendants of both families held key administrative positions in the province.

The high social status of land surveyors is still visible in late-second-millennium Kassite Babylonia thanks to ***kudurru*** stones recording royal grants of land which could be measured in seeds necessary to sow the field. The measurements were done by a land surveyor and supervised by local and royal authorities (Robson 2008a: 167–76). A dozen land surveyors attested on these inscriptions are said to be descendants of Arad-Ea, who lived two or three centuries earlier. Arad-Ea, himself an expert in accountancy, was the son of the governor of Nippur. These land surveyors thus belonged to the lineage of a very powerful and rich family which included royal scribes and high officials.

As an echo of the tasks performed by land surveyors, field plans from the late third millennium as well as mathematical texts from the Old Babylonian and perhaps also the Kassite periods concern the computation of the areas of fields of different shapes (Nemet-Nejat 1993: 72–8).

Egyptian Land Surveyor

The land surveyor profession is presumably as ancient in Egypt and might be attested by autobiographic hieroglyphic inscriptions found in the graves of high officials working in close relation to the king. This could be the case for Metjen, whose tomb (Mastaba L6) excavated at Saqqara, north to the Djoser's pyramid, dates to the IVth Dynasty (27th century, Strudwick 2005: 192–4). The biography of Metjen, one of the earliest written on a wall of a tomb, gives a good insight into his life and position in society (Allen 2013: 208–13). Metjen presumably started as a simple scribe, overseer of a provision bureau, a position inherited from his father, himself an honorary scribe. He became a high official of the Old Kingdom and provincial governor, ruling the eastern part of the Fayyum. For all his services to the pharaoh, he received by royal decree twelve domains with their peasants in three provinces of the Nile Delta as well as a large orchard land from the royal domain and a huge, well-equipped house with its personnel. The IVth decree gives the areas of the fields suggesting the ability, at that time, to compute areas of geometrical figures (Imhausen 2016: 48). Among his titles, Metjen held the office of "Crier," in charge of evaluating the quantity and value of produce and crying out the figures to the recording scribes (Clagett 1999: 146). Metjen was a scribe, able to compute, and whom, in the service of the king, has risen above his father's social level.

Another example comes from the tomb of Menna at Thebes dating to the fourteenth century. This high official, married to a noblewoman, was both

FIGURE 2.1. Wall painting from the tomb of Menna in the Valley of the Kings. Rogers Fund, 1930, The Metropolitan Museum of Art.

referred to as a scribe and the overseer of the palace and temple agricultural estate. A painting represents him, with a group of people, measuring the size of a field with ropes and computing the respective number of products to deliver (Figure 2.1; Hartwig 2013).

A wisdom literary composition (BM 10474), *Instructions of Amenemope*, dating to the last centuries of the second millennium, contains advice to his son on how to behave in times of crises to have a successful life. In the introduction, we learn that Amenemope was responsible for the borders of fields and thus measured land. For this, he used the measuring rope. He also had to control the standard capacity vessel and the measuring of grain (Imhausen 2016: 158–60).

Calculating the surface of geometrical areas is well documented in the mathematical problems proposed, for example, by the Rhind Papyrus. The different geometrical shapes considered, rectangles, triangles, trapezoids—except the round one—could correspond to fields.

Accounts of an Egyptian Ka-Servant

Computations in the form accounts could also be carried by Egyptians of the lower levels of the landed gentry. A group of papyri found in a tomb complex at Thebes documents the life and activities of a middle-class man, Heqanakht, a *ka*-servant (or *ka*-priest) during the first half of the twentieth century (XII Dynasty; Allen 2002). The *ka*-servant was paid by the vizier Ipi to perform the daily offerings at his tomb. Heqanakht, whose household of some twenty people, including a number of servants, was in another city, and had religious and administrative duties: besides his own lands, he had to manage lands and personnel located in the Thinite nome, providing income for the funerary cult of Ipi, who had organized in advance the perpetuation of his mortuary cult. Heqanakht left letters and accounts dealing with land cultivation and loans; as a *ka*-servant, he was literate and numerate, and has presumably written himself several of these texts. Heqanakht had to estimate the yield of the land, proceed to grain measurements, compute, and distribute rations according to rates between different types of grains. He computed values in grain of different commodities as oil, textiles, or copper. Some of his accounting activities, using operations on fractions, recall problems presented in the mathematical papyri as, for example, the quantity of grain necessary to produce bread.

Old Assyrian Merchants

Besides officials working for the administration, some merchants, acting privately, had to perform computations on a daily basis as part of their activities. The earliest and best documented private merchants are the inhabitants of Aššur (Iraq), who developed a long-distance trade in Central Anatolia and settled in some forty different towns there during the first centuries of the second millennium. The center of the Assyrians' operations was located at Kaneš (Kültepe, near Kayseri), where 23,000 cuneiform tablets have been unearthed (Figure 2.2; Larsen 2015).

During their transactions, from the purchase of merchandise in Aššur until their sale in Anatolia, or during their commercial partnerships, the merchants had to realize various computations (Michel 2008). They were using the Mesopotamian weight system to quantify the metals, objects of their commercial exchanges. Weighed silver was used as a mean of payment. They carried out weighing taking into account deficient weights. In order to compute the value of a given metal into silver, they had to multiply or to divide the quantity of metal by the specified conversion ratio (Text 2.1; Michel 2020a). They built up various types of contractual relationships to gather important amounts of silver or to increase the capitals already obtained. They were involved in commercial loans and needed to know how to calculate the due interest. Their trade was built on joint-stock companies, in which one or several members put an amount of silver or gold at the disposal of an agent to perform commercial transactions. At the end of the operations, the partners set their accounts, paid their debts, and shared the profits according to certain rules, all of which led to complex computations. A large part of Aššur's population seems to have been involved in this long-distance trade, including the king and his family, the high officials, priests and temples, and the merchants' families, both men and women. School tablets show exercises which correspond perfectly to the knowledge merchant needed to master (see below).

FIGURE 2.2. Photo of TC2, 24 an extract of a letter concerning transactions in tin. Photo: C. Michel, courtesy of the Department of Oriental Antiquities at the Louvre.

Egyptian and Mesopotamian Architects

Egyptian monumental buildings, and especially pyramids and tombs as early as the first half of the third millennium (III–IV Dynasties), have acquired a universal reputation for their main building material, the stone that has stood the test of time. These came from carriers in which inscriptions give an idea of the number of workers who were employed to extract a specific quantity of blocs. Some of the Hieratic mathematical papyri propose problems linked to construction planning, with the computation of the number of workers needed to complete a task linked to an architectural project (Imhausen 2016: 112–24; Monnier 2019). Egyptian architects of this period invented both new construction techniques and new geometrical concepts.

To the 2 *gu* 18 *mana* of tin of your consignment 13 5/6 *mana* of tin accrued for you as road tax;

3 *mana* 15 *gin*: food for 2 donkeys and the harnesser;

1 donkey died and imposed an expense of 17 *gin* of silver;

7 *gin* of silver cost the hire of a donkey driver from Hahhum to Kaneš;

12 *gin* loss on the tin for current expenses was deducted from the 12 *mana* of tin for current expenses;

you will have to balance 5 *mana* 15 *gin* of tin on the tin for current expenses, at 8 *gin* each (*gin* of silver), its price in silver: ½ *mana* 9 1/3 *gin*;

I have added 17 *gin* the price of a donkey, and 7 *gin*, the hire of a donkey driver, so in all: you shall balance 1 *mana* 3 1/3 *gin* of silver;

2 ¼ *gin* of silver, which they will balance for your, has been credited, so I have balanced only 1 *mana* 1 *gin* 15 *še* of silver with Aššur-malik

Weights: *gu* ←×60– *mana* ←×60– *gin* ←×180– *še*

1 *gin* ≈ 8.3g

TEXT 2.1. Extract of a letter concerning transactions in tin.

Among the very few written testimonies of these innovations is a limestone ostracon found in the complex of pharaoh Djoser, at Saqqara, showing the profile of an elliptic vault in red ink with some measures written in Hieratic (Figure 2.3; Rossi 2003: 115–17; Lightbody and Monnier 2017). The author of this diagram is anonymous. By coincidence, the most famous Egyptian architect who would have designed and supervised the construction of the funerary complex of Djoser, with its step pyramid, at Saqqara, Imhotep, is contemporaneous of this ostracon. As for the heroes, between legendary figure and historical celebrity, a lot of talent has been attributed to Imhotep, first as sculptor and architect, but also as vizier of the king, medical doctor, etc. Egyptians revered him as a son of divinities and praised his talents for more than two millennia.

Another high-ranking architect from this early period is Hemiunu (twenty-sixth century), the grandson of the IVth Dynasty's founder. He built the great Giza step pyramid. An inscription gives him as head of the judges and vizier of the pharaoh. A statue showing a wealthy man, representing Hemiunu, was found in his mastaba in the Giza necropolis. There is no doubt that many scribes

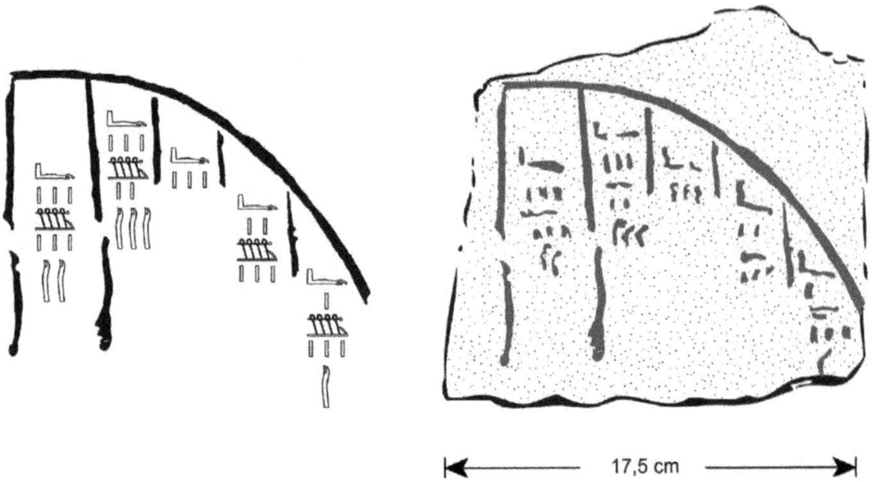

FIGURE 2.3. Drawing of the ostracon JE 50036 (Lightbody and Monnier 2017). Drawing courtesy of Franck Monnier.

were working for these architects, and presumably produced some of the calculations, but they remain anonymous.

Mesopotamia also provided annotated building plans, some of them are detailed enough to represent each brick of a paved courtyard. The only one that gives both the names of the rooms of the building and their dimensions was excavated at Tello, ancient Girsu, and dates from the end of the third millennium (Figure 2.4a–b). If the measurements are correct, the plan does not respect the proportions (Gruber and Roaf 2016). Unfortunately, such plans are never signed and so we do not know the identity of their authors.

Most of the buildings in ancient Mesopotamia were made in mud bricks, including the great ziggurats, multistory towers with a religious function. Administrative tablets from the late third millennium provide important details about the construction of the ziggurats and the number of workers to employ. These activities were based on standard workdays setting the quantities of bricks shaped and transported. Thus, to build only the first terrace of the Ur ziggurat a thousand of workers would have work for more than five months just to prepare and transport the 7.6 million bricks needed (Sauvage 1998). Workers were grouped in teams under the responsibility of supervisors themselves under the direction of a mason in charge of the reception of the material and brick laying. As the construction was officially the great work of kings, the names of their architects who draw the plans and directed the work remain unknown. Tablets give the names of the intermediaries in charge of the organization of the work.

Many problems presented in the Babylonian mathematical texts deal with bricks and brick works; they address some questions an administrator would need to answer, such as the quantity of earth transported and the number of workers to employ (Nemet-Nejat 1993: 27–35, 43–55; Robson 1999: 57–92). The dimensions of the bricks used in mathematical texts correspond to those found in texts of practice as well as in the excavations of Mesopotamian sites (Sauvage 2020).

FIGURE 2.4a. Photo of the tablet AO 338, Louvre Museum. Photo: C. Michel, courtesy of the Department of Oriental Antiquities at the Louvre.

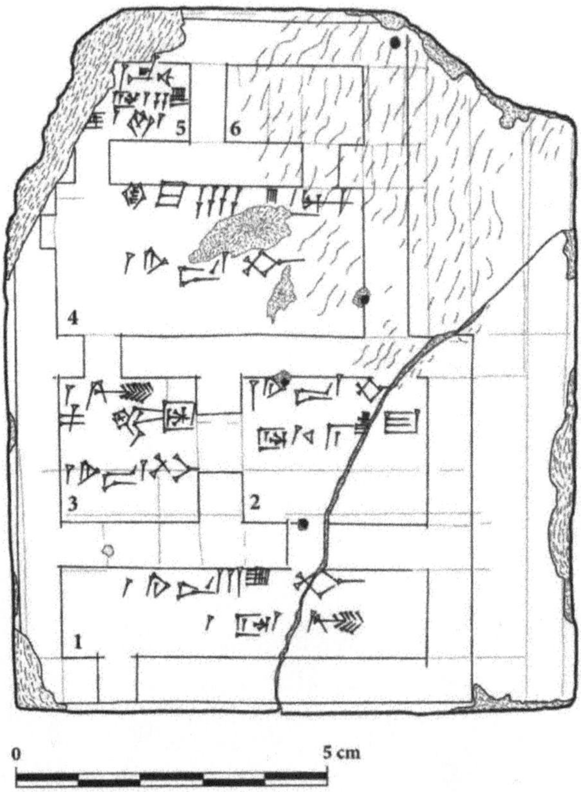

FIGURE 2.4b. Drawing of AO 338 by Gruber and Roaf (2016). Drawing courtesy of Martin Gruber.

AUTHORS AND OWNERS OF MATHEMATICAL TEXTS

As already noted, land surveying, architecture, management, and trade are among the themes which inspired the problems presented in the Old Babylonian mathematical texts. The quantities mentioned in mathematical texts were often concrete, thus the mathematics taught was often rooted in daily life. The issue at stake now concerns the writers and the authors of these mathematical texts. While authorship is never neglected in Greek mathematics, as we shall see in Chapter 3 below, mathematical texts from Mesopotamia and Egypt are, for the most part, anonymous. When, by chance, the name of the writer is given, there are rarely indications as to his status and biography. Moreover, the writer is not necessarily the author of the text, he may have copied it from another manuscript. Some mathematical texts have been found in context, and it is possible to shape the profile of their owner. However, there is no proof that the owner of a mathematical text and its author are the same person. Despite all of these difficulties, this section intends to draw a portrait of the writers, authors, and owners of mathematical texts in Mesopotamia and in Egypt and to show that they emanate from different communities of scribes.

Notables and Scholars of Third Millennium Mesopotamia

There are relatively few texts for the third millennium that may be considered as “mathematical texts” by contrast to texts of practice involving mathematical computations. They are usually connected to large scale works and the control of land by high ranked officials. Among these, a group of texts from southern Mesopotamia dating to middle of the third millennium are linked to the quantification of surfaces. These texts present tables linking the sides and the surfaces of rectangles or squares. Even though they may refer to land-surveying practices, the arrangement of data which separate numerical and metrological ones testify to the introduction of new mathematical concepts (Chapter 3). The earliest examples have been produced by the scholarly milieu of Šuruppak together with lexical lists and Sumerian literary texts. Their authors are learned scribes who used to add their names in a colophon at the end of the tablet; but few are complete. Another example, from an unknown context in Adab, contains a table of increasing sides of squares and their corresponding surfaces. This tablet has been written by the scribe Nammah, whose name was very popular at that time (Edzard 1968). Another text from Zabalam, which shows an attempt to compute small surfaces by creating subunits of the smaller area unit *sar*, is authored by Lugal-hegal-su. Once more, there is no data about this scribe. According to Proust (2020), these tablets emanate from an erudite scribal milieu; the mathematical methods developed there have not affected the contemporary administration practices.

For the Sargonic period which saw the first centralized state in Mesopotamia (24–23 centuries), there are a dozen mathematical texts presenting problems, all of these coming from the south, and not from the Akkadian-speaking north. Some of these, focusing on the computation of areas of squares and rectangles, have been found in a provincial town, nearby Lagaš. They could have belonged to a notable, perhaps the governor of that town, who would have kept his tablets produced during his scribal education (Foster 2016: 216–19). Other mathematical texts deal with irregular figures, presumably parcels of land, for which the scribe has to find the area or one side of the geometrical shape. The resolution of these problems suggests the conversion of metrological measurement units into numbers using a positional system and computations carried on an abacus. Two of these problem tablets have been written by Ur-Ištaran, perhaps administrator of the Girsu palace (Foster 2018: L. 4552). If not dealing with a homonym, then it would confirm the link between mathematical texts and high officials.

Students and Masters in Early Second Millennium Mesopotamia

The largest number of mathematical texts date to the Old Babylonian period. They have been found in a context of education in Nippur, Sippar and Ur (south of Iraq), in several sites of the Diyala Valley, in Mari (Syria), and in Susa (Iran). They reflect different traditions with local specific features. Hundreds of such texts found in several houses at Nippur, allowed a reconstitution of the scribal curriculum in this main cultural center. Unfortunately, there is too little data about the owners of these houses.

A house excavated at Tell ed-Dêr is one of the few to allow drawing the portrait of one or two authors of mathematical school texts. It belonged successively to two different persons and contained some 2,000 cuneiform tablets, including eighty linked to a teaching activity (Tanret 2002). The main archive belonged to the family of the great lament priest Inanna-mansum of the city of Sippar-Amnānum. He chose his eldest son, Bēlānum, as successor in his religious function. The sixty-eight school tablets found in his house were all written by this son who became lament priest under the name Ur-Utu in 1642 after his father's death. Among these were six metrological tables for capacities and weights and two mathematical exercises focusing on the computation of field areas (Tanret 2002: nos 42–7, 55–6). Bēlānum learned basic mathematics, allowing him to carry mathematical operations for daily economic life. His teacher, Šumum-liṣi, was a professional scribe working for the family. He wrote administrative documents, contracts, and letters and authored 151 texts found in the house. Bēlānum was not destined to be a scribe, but to have a high responsibility in the temple with an important social status. However, he had to be able to check what was written for him by scribes, and this explains his basic education; his master provided him a passive knowledge.

Before the house was rebuilt and occupied by Inanna-mansum family, it belonged to Lamassāni, daughter of Rīš-Šamaš, a *nadītum*-priestess of the Sun God. These priestesses, who belong to the urban elite, were forbidden to marry and reproduce, but they could take care of a niece also destined to become a priestess. In the house of Lamassāni, twelve school tablets were discovered, two of them linked to a mathematical education: a metrological list and a multiplication exercise with reciprocals. These could have been written by the niece of Lamassāni during her education (Tanret 2002: nos 57 and 59: 140).

This interesting remark raises the issue of literate and numerate women. The archives of the *nadītum* of Sippar have proven the existence of some twenty female scribes working for the priestesses, some of them being themselves *nadītum*. However, education was not limited to the religious sphere. One of them, Inanna-ama-mu, was the scribe of the judges of the Sun God temple. These female scribes did not work exclusively for women and some male scribes also wrote the same types of documents for the priestesses: field purchases, leases of land, inheritances, donations, etc. To confirm the existence of female scribes at Sippar, four school exercises containing a colophon were written by a female apprentice scribe; they followed the same curriculum as schoolboys (Lion 2011: 99–100). One of these women is munus-dumu-é-dub-ba-a, a Sumerian title referring to a female scribe who would have gone through the whole curriculum and became a female Master (Michalowski 2012). There are, up to now, no mathematical texts authored by a woman, however it is very likely that female scribes had received the same education in mathematics as male scribes.

The mathematical school tablets found at Sippar belonged to the elementary level. By contrast, advanced problems may have been written by high level students, by masters or by scholars. These texts were not all written for the purpose of education. Several procedure texts from South Mesopotamia present lists of solved problems of progressive difficulty linked to practical problems, their resolution is based on the arithmetical tools scribes learned during the elementary level. These texts are drawn by masters for advanced teaching. Linked to these procedure texts are catalogues compiling the problems without giving the procedure, thus archiving educational texts (Proust 2014: 87). Other long lists of problem statements presented in an arborescent structure were not intended to be solved. They are the result of a reflection on mathematic writing techniques by scholars not linked to education but simply interested by developing new mathematics. The houses in which they were presumably preserved, together with elementary mathematical exercises and tables, were both educational and intellectual centers. Some of these mathematical texts could end with a colophon including the date, a proper name, and data linked to the text itself as keyword, and sections. Unfortunately, as they come from illicit excavations, the profile of the scholars who produced these mathematical catalogues and series remain unknown.

Scholars of Late First Millennium Babylonia

A few mathematical texts date to the Achaemenid (539–330) and Hellenistic (330–140) Babylonia, but many more astronomical tablets testify to the use of advanced mathematics (Ossendrijver 2012). At that time, the main spoken languages were Aramaic and Greek which were written with ink on papyri, only scholarly works were still written in cuneiform on fresh clay. The authors of the mathematical tablets found at Uruk and Babylon, are known thanks to the colophons in which the scribes mentioned the writer of the text, the place it was composed, its content, eventually the original document from which it was copied, and the owner of the tablet. They belong to important lineages of scribes that we can reconstruct over several generations, the most learned were referred to as "masters of knowledge" or "experts" (*ummānu*). These men with an encyclopedic culture had students, among whom their own sons, who were writing tablets for them. These mathematical tablets belonged to families involved in medicine, astrology and astronomy.

A dozen mathematical texts dating to the Achaemenid period and showing different methods to evaluate areas were discovered in the house of the exorcist-doctors, *āšipu*, at Uruk (Proust 2019), together with forty astrological and astronomical tablets (Steele 2019a), contracts and a great variety of scholarly texts: literary compositions, hymns, omens, exorcisms, medical recipes, etc. (Clancier 2014). Many belonged to two families over respectively four and three generations: the Ekur-zakir descendants succeeded to those of Šangu-Ninurta after rebuilding the house (Proust and Steel 2019a: 8–34). These tablets are linked to the education of the younger members of the family, to the practice of the profession of exorcist-doctor, to the encyclopedic knowledge of the senior family members as well as their duties towards the temple. They confirm that important professions as well as scholarly knowledge, including mathematics, was transmitted from father to sons over several generations.

Another group of mathematical texts originates from the library of the Reš temple at Uruk and belonged to priests and astronomers. As well, these texts were found together with a great variety of scholarly tablets including astronomical texts using advance mathematics. Among the writers of these mathematical tablets, we find both experts and younger scholars copying texts for the former. As he was a student, Anu-aba-utēr wrote four astronomical tablets for Šamaš-ēṭir. The latter was the great priest of the Reš temple and an astronomer. Anu-aba-utēr, also known as astronomer, was the owner of a mathematical text containing seventeen problems on arithmetical, geometrical and metrological topics (Ossendrijver 2019). He wrote seven tablets for his father, including a Sumerian cult song, and several computational astronomic texts (Robson 2007b). At that time, a young scribe studied within his family, having learned from his father or uncle and copying texts for him, but he also completed his education with a master of another literate and numerate family. Contracts found together with the scholarly tablets suggest that some of these families were linked by marriage. Most of these scholars were priests or working for the temple; they put their astronomical knowledge to the service of the temple, predicting astral phenomena based on the compilation of observations, data quantification and development of mathematical procedures in order to build the astronomical tables with the recurrence and periodicity of the observed phenomena (Ossendrijver 2012).

During the middle of the fourth century the Esagil temple of Babylon also employed astronomers, a profession transmitted from father to son and for which they receive a salary of one *mana* of silver (500 g) per year and a cultivable land. They cumulated this office of astronomer with that of priest or exorcist-doctor. During the last centuries of the first millennium, under the Greek domination, the Babylonian temples, with their libraries and experts, represented the centers of knowledge where all innovations in mathematics, astronomy, and medicine were made (Beaulieu 2006).

Writers of the Egyptian Mathematical Texts

The question of authorship of the Egyptian mathematical texts is extremely difficult to answer because of their fragmentary state. Written on a perishable medium, they survived in small pieces that had to be reassembled; their colophon with the name of their writer may be missing. Moreover, several mathematical papyri come from the antiquity market and their context of discovery is lost. When the name of the writer is known, his life and social status might not be documented; moreover, the writer may have simply copied the mathematical texts from an earlier manuscript.

This is the case, for example, of the Rhind Papyrus (32 × 500 cm), written in Hieratic, presumably the most well-known sample of Egyptian mathematics. It was bought by A. Henry Rhind and is supposed to come from a small building nearby the Ramesseum (mortuary temple of Ramses II) at Thebes (Clagett 1999: 113). The mathematical text, on the recto, dating to the Second Intermediate Period of Egypt (*c.* 1550), was copied by the scribe Ahmes from a lost original, presumably dating to the second half of the nineteenth century. The text starts as shown in Text 2.2.

It gives then tables and series of problems supposed to be designed for educational matter. One third of the papyrus contains tables for the division by 2 of integers, and various problems involving divisions and multiplications of fractional expressions. Then only, problems are linked to applied mathematics with divisions of a measure, of beer and bread for rations, computing volumes of cylindrical and rectangular granaries, areas of different shapes and ending with the slopes of pyramids, and miscellaneous algebraic and geometric problems. As suggested by the introduction of the papyrus, an Egyptian official scribe was expected to be able to cope with such problems in the performance of his activities (Imhausen 2016: 65–7).

> Rules for reckoning, for inquiring into things, and the knowledge of all things, mysteries to all secrets. This book was copied in regnal year 33, month 4 of Akhet, under the majesty of the King of Upper and Lower Egypt, Awserre, given life, from an ancient copy made in the time of the King of Upper and Lower Egypt, Nymatre. The scribe Ahmes writes this copy.

TEXT 2.2. Beginning of the Rhind Papyrus, translation Clagett (1999: 122).

A quite atypical text on the verso of the papyrus is a historical and political chronicle written in the Hyksos context (Oriental Nile Delta), a kind of marginal milieu in Egypt. It gives objective and factual dates and events that led to the fall of Avaris in an exceptional way as it expresses the point of view of the defeated party (Barbotin 2008: 180–1). The papyrus was later taken to Thebes after the victory of this city. The texts on both sides seem to have been written in a short time; it is not impossible that Ahmes himself wrote or copied this chronicle.

The Rhind Papyrus is the only mathematical Hieratic text bearing the name of its writer. It has been suggested that most of the mathematical papyri have been composed during the XIIth or early XIIIth Dynasties, i.e. during the first half of the second millennium (Clagett 1999: 113–280). Thus, they are contemporaneous to the bulk of the cuneiform mathematical texts. They originate from different sites but their exact location remains most of the time unknown; when it is known, they were found in a tomb:

- Thebes for both the mathematical leather roll (Imhausen 2016: 69–70) and the Moscow Papyrus (Clagett 1999: 205–6),
- tomb N 408 of This necropolis for the Reisner Papyrus (Clagett 1999: 261–2),
- the workers' town Ptolemais Hormou (El-Lahun), nearby the pyramid of Sesostris II for the Lahun fragments (Clagett 1999: 239–40),
- Akhmim for the Cairo wooden board (Imhausen 2016: 68).

The tomb N 408 in the This necropolis belonged to a high official of the city. Together with the Reisner Papyrus were four rolls of papyri with administrative texts linked to the management of resources at the royal dockyards, the construction of a building, and a list of distribution of rations (Hagen 2018: 75).

The tables used to compute with fractions and the most complete texts with series of problems and their solution suggest that they were used as school textbooks in the same manner as the tables and problems addressed to Babylonian apprentice scribes, but the resolutions used different methods (Couchoud 1993: 4–5; Bernard, Proust and Ross 2014). According to Annette Imhausen (2003), half of the proposed problems refer to practical matters as computation of rations, volume of granaries, field areas, and other problems linked to boats, brewery, breeding, etc.; they might have been part of the education of scribes working for the administration.

The group of Demotic mathematical papyri has been less studied (third century BCE–second century CE; Parker 1972). Some contain also other types of texts, written at about the same time or showing traces of reuse. Several of them come from the antiquity market, others have been found in temple libraries at Tebtynis and Soknopaiou Nesos, and the last two in the Temple of Tutu at Kellis.[6] The longest of these papyri, which includes mathematics from simple exercises to the computation of the volume of a pyramid, is written on the verso of the *Code of Hermopolis* (first half of the third century) which deals with the relationships between owner and tenant, problems of property, marriage, etc. with some private commentaries. Mark Depauw (1997: 103–5) suggested that the text is a legal and mathematical reference book. There is no indication about its writer, who might have been a scholar, knowledgeable in several disciplines.

MATHEMATICAL PREPARATION FOR PROFESSIONS

Both in Egypt and Mesopotamia, mathematical training is part of the education of scribes. This profession, considered as the best among all, is reserved to privileged individuals; the intellectual work of the scribe is regularly opposed to tiring manual professions. However, to become a scribe, one had to study hard for several years. Literary compositions about school survived from both ancient Mesopotamia and ancient Egypt. They depict the organization of the education and the topics that were studied, including metrology and mathematics. Texts produced in the context of education provide data about the content of the lessons from elementary to advanced level. They show a great variety in the practice of mathematics and in its transmission according to the milieus considered. In Mesopotamia, the mathematics taught at Nippur (eighteenth century) to the students of the upper society was completely different from that studied by the future Assyrian merchants in Aššur and Kaneš. For ancient Egypt, the sources are scarcer. The advanced level of education, which could last up to twelve years, depended on the future career of the students, whether in administration, priesthood (in the House of Life), or military.

The Edubba Compositions and the Social Profile of the Scribes

For the twenty-first century, hundreds of thousands of texts document the management of royal goods which caused multiple controls and an increased production of tablets. The number of scribes dedicated to administrative tasks increases dramatically: 1,600 of them are known by their names. There is a great need of well-trained scribes, and the literate King Šulgi claimed in a hymn to have set up schools in many cities of his country (Text 2.3).

The "House of the Wisdom" refers to schools which were part of temples, especially those dedicated to Nisaba, the goddess of the scribes during the Ur III Empire. Such a tradition could have survived during the Old Babylonian period in parallel to the education in private houses (Charpin 2017: 110–15). In this

> When I was small, I was at the Edubba, where I learned the scribal art from the tablets of Sumer and Akkad. None of the nobles could write on clay as I could. There where people regularly went for tutelage in the scribal art, I qualified fully in subtraction, addition, reckoning and accounting. (. . .)
>
> In the south, in Urim, I caused a "House of the Wisdom" of Nisaba to spring up in sacrosanct ground for the writing of my hymns; up country in Nippur I established another. May the scribe be on duty there and transcribe with his hand the prayers which I instituted in the E-kur; and may the singer perform, reciting from the text. The Edubba are never to be altered; the places of learning shall never cease to exist.

TEXT 2.3. Hymn Šulgi B. Extracts from lines 10–20, 308–19 (trans.: http://etcsl.orinst.ox.ac.uk/section2/tr24202.htm, accessed September 20, 2019).

hymn, the Edubba, the "House of the Tablets," i.e. "the school," appears as a place where nobles and princes learned writing, computing, and accounting.

The organization of scribal education in the late third millennium southern Mesopotamia might be reflected by the Sumerian "Edubba compositions" written and copied during the first centuries of the second millennium. These texts provide a social profile of the students going to school: most of them came from wealthy families of the social elite. At home, the schoolboy recited his tablet to his father assuming that the father was literate and able to control homework, if not a professional scribe, he might have had an intellectual activity (Attinger 2017). In another composition, a father recommended his son to be attentive at school in order to be able to succeed him in his activity; this is a good example of social reproduction (Text 2.4).

The Edubba compositions depict the education of the young scribe in schools: the life of the schoolboy, his relations with his master, his friends, and his parents. They advocate the learning of the Sumerian language and give interesting insights into the curriculum, including mathematics (Text 2.5).

The education was based on the copy and the memorization of long lists of items: signs, words, personal names, professions, etc. The apprentice-scribe also

> Of all the human professions which exist on earth and of which the god Enlil has named, he has not named any profession more difficult than the art of the scribe. In accordance with the prescriptions of the god Enlil, the son must succeed his father in his profession.

TEXT 2.4. Edubba B, Sjöberg (1973).

> Now I've been here for the stipulated time, I can cope with Sumerian,
> scribal work, archiving, accounting, calculation! (. . .)
> I want to write tablets:
> The tablet of measures from 1 *gur* of barely to 600 *gur*,
> The tablet of weights from 1 shekel to 20 minas of silver,
> With the marriage contracts that people will bring me,
> The partnership contracts, I can choose controlled weights of one talent,
> The sale of houses, of fields, of slaves,
> The guaranties in silver, contracts of field rental,
> Contracts of cultivation of palm date grove,
> Even adoption contracts, I know how to write all this.

TEXT 2.5. Edubba D, translation Civil (1985).

learned the metrological systems by copying metrological lists, in order to be able to control the weights and proceed to the computation of balance sheets: all knowledge important for goods management and administrative tasks.

Curriculum at Nippur

Because they are literary compositions meant to entertain the reader, the data given in the Edubba compositions must be treated with caution, but they give hints about the main disciplines taught, Sumerian and mathematics, in a presumably late-third-millennium context. There are great changes in the educational system at the very beginning of the second millennium, with a growing emphasis on writing. Before, lists and tables were perhaps transmitted orally (Proust 2015).

The lists mentioned in the Edubba texts correspond to those found in several Old Babylonian sites, especially at Nippur in some houses where a master taught a few students. In the 45 m^2 "House F" archaeologists found 1,300 school tablets dating to the first half of the eighteenth century, 90 per cent being school material recycled as construction material in the walls, and in a clay basin. The house served as a place of education, but the tablets were later discarded (Robson 2001b).

School tablets show different shapes and layout that have been arranged in four groups corresponding to different functions (Civil, Green, and Lambert 1979). Type II tablets have different exercises on the obverse and on the reverse and can bear two hands: the master's and the pupil's copy. Niek C. Veldhuis (1997) observed that one side presents a new lesson while the other contains a lesson learned previously; he could thus reconstruct the progression of the items taught and the curriculum of the scribal education at Nippur, distinguishing different phases. Applied to mathematics, this method allowed establishing the general structure of the mathematical training curriculum (Proust 2007; Robson 2009); it was articulated with that of writing and Sumerian (Proust 2014).

During the elementary level, the pupil learned how to draw signs and learned the vocabulary organized by themes. In parallel, he learned by rote metrological lists enumerating measurements of capacity, weight, surface, and length. At the end of the elementary level, the list to be memorized were more complex with signs arranged by their shape, metrological tables giving the correspondence between the measurement units and abstract numbers written in sexagesimal place-value notation (SPVN, Proust 2010), and numerical tables: reciprocals, multiplications, squares, square roots, and cube roots.

During the intermediary level, the pupils copied proverbs, contract models in Sumerian, and had to solve small mathematical exercises based on operations with SPVN numbers. These were linked to the computation of areas of various geometrical figures. During the advanced level, they copied Sumerian literary compositions and solved mathematical problems using all what they had learned before (Proust 2014; Bernard, Proust, and Ross 2014). The school tablets found at Nippur present a great homogeneity by contrast to what was found in many other cities where the typology of tablets varies a lot, suggesting a different organization.

Mathematical Training through Practice: The Case of the Assyrian Merchants

This is the case for many of the "school tablets" found in the houses of the Assyrian merchants at Aššur and Kaneš. Besides large rectangular tablets with lists of weights given in an increasing order, words, personal names, and sentences extracted from letters, a dozen round small tablets concern exclusively arithmetic exercises. Most of these consist in converting a quantity of a metal into another metal according to a given conversion ratio.

In Text 2.6, the scribe had to calculate the weight of silver necessary for the purchase of copper which is quantified by a top-pack-*elitum*, knowing that a top-pack-*elitum* weighs 30 *mana*; he failed since the result is not correct. The computation was presumably done by a rule of three, which might have been part of the arithmetic training of the apprentice merchant.

2 *gu* 14 *mana* of copper

the top-pack-*elitum* at 1/2 *mana*

5 *gin*. Its (value in) silver: 2 *mana*

6 1/2 *gin* 15 *še*

[The correct result is 2 1/2 *mana* 6 1/3 *gin*]

TEXT 2.6. Old Assyrian school tablet from Aššur. *Comments*: Michel (2020a, text 2.1F).

To solve such exercises, the Assyrians did not convert their quantities in the SPVN system in use at Nippur, but carried their computations on the metrological values, including fractions, starting with the largest measurement units (Michel 2020a). These merchants received their mathematical education through practical professional tasks. They were taught by a master or by a family member. Their wives and daughters were fully involved in the long-distance trade, earning money from the sale of their textiles in Anatolia and investing their revenues in financial operations. Many were numerate, keeping their accounts and claiming the amounts due to them. Some women, too, have left tablets. Like men, they presumably learned the basics of writing and computing from their parents (Michel 2020b).

The Transmission of Mathematics in Egypt

The transmission of mathematical knowledge in Egypt is less documented due to the lack of school exercises by apprentice scribes.[4] However, as for Mesopotamia, literary compositions document the education of scribes, a profession which is also considered as one of the most important in the Egyptian society. In a letter to his son who left school to work in a field a man depicts the hard manual work he will have to face and how he will be put under pressure by the scribes who calculate the taxes he will have to pay at the time of harvest. By contrast, the "scribe he is the taskmaster of everyone. There is no taxing of the work of the scribe. He does not have dues" (Anastasi V, 17, 2–3, Ramesside period, Simpson 2003: 439). The scribe is the one to compute the taxes due by the others.

In another section of the papyrus, a father explains to his son the benefits of going to school and the topics that will be taught to him (Text 2.7).

The school mentioned here could allude to the "House of instruction," a royal school already known from late third and early second millennia texts, thus contemporaneous to the bulk of the mathematical papyri. In the temples, scribes were linked to the "House of Life" where they learned religion, but also mathematics, astronomy, and medicine (Bernard, Proust, and Ross 2014:

> I have placed you at the school along with the children of magistrates in order to instruct and to teach you concerning this aggrandizing calling (. . .) You bring your book daily with a purpose, be not idle. They say, "Three plus three." You . . . another happy occasion, in which you grasp the meaning of a papyrus book (. . .) You will make calculations quietly: let no sound coming forth from your mouth be heard (. . .) Fall in with the ways of your instructor and hear his teachings. Be a scribe.

TEXT 2.7. Papyrus Anastasi V, 22, 6–7, 23,1–2, 4–6 (Imhausen 2016: 143–4).

36–8). The text gives hints on the mathematics learned, additions and computations based on a book, presumably those presenting problems to solve which have been referred to as mathematical papyri above and were addressed to male students only. These texts reflect an advanced level in mathematics; there are almost no text documenting an elementary level of mathematical education.

A satirical letter, Papyrus Anastasi I, dating to the nineteenth Dynasty (*c.* 1250), was sent by a military scribe, Hori, to his colleague scribe Amenemope (Mapu), accusing him of not meeting the standards of the profession (Couchoud 1993: 183–4; Imhausen 2016: 148–56). He blames him for being incompetent in terms of mathematical and technical matters despite his high rank in military administration. To illustrate his accusations, he gives examples which concern the resolution of mathematical problems that he considers Mapu unable to solve (Anastasi I, 13,8–18,2): computing the number of bricks needed to build a ramp of given dimensions, the number of workers needed to move an obelisk which dimensions are specified, the number of workers to erect a colossal statue using a box system filled with sand, or the rations of a military group of a given number of soldiers. Three of these problems focus on the number of workers necessary to complete a task and their cost in terms of wages. This echoes the many mathematical problems found both in Egypt and Mesopotamia.

These scribes worked for the army and had different military ranks. More generally, scribes had a variety of social positions according to the domain in which they work and their tasks. The ordinary scribes who produced the administrative and economic documents might have been in charge of daily accounting, using the skills they learned at school. Working for temples or for the state, they did not belong to the upper class (Imhausen 2003). By contrast, royal scribes, as high officials, belonged to the elite and could pay to have their personal decorated tomb.

CONCLUSION: A DIVERSITY OF USES IN A VARIETY OF MILIEUS

This chapter has identified a variety of practitioners of mathematics and their activities through the texts discovered in Mesopotamia and Egypt. Mathematical texts and texts documenting mathematical practices include different mathematical traditions developed for different goals in different milieus. Mathematical texts appeared together and developed in parallel with other scientific texts, including those in the fields of divination, astronomy, and medicine. At the beginning of the second millennium, both in Egypt and Mesopotamia, we observe a very fertile production of mathematical problems rooted in daily life activities. To solve them, different sophisticated mathematical methods were developed by highly educated Egyptian and Mesopotamian scribes and masters. A variety of mathematical cultures existed also within these two geographical areas. Although this chapter considered the ancient Near East and more precisely Mesopotamia and Egypt, no claim is being made that these mathematical cultures, which already show a great diversity, are representatives to all others, especially Greek and Roman, not to speak of China and India.

Many of the mathematical texts with problems are composite, they represent a compilation of some earlier texts. They were regularly copied over several centuries, whether for educational purposes or for patrimonial purposes, in order to be kept in libraries. As for texts indicating other kinds of mathematical practices, one must distinguish the many agents involved in producing them, including high officials, royal scribes, heads of bureaus, administrators, and accountants.

In both Mesopotamia and Egypt, there were no specialists in mathematics: there were scholars who went through the complete curriculum and were knowledgeable in various domains including mathematics, astronomy, medicine, literature, etc. Mathematics was a subject taught in most "schools," but with variations in its content. For example, the mathematics studied in Nippur schools and in Aššur merchants' houses was quite different in the topics covered and its procedures. In Hellenistic Mesopotamia, by contrast, the general and mathematical knowledge of scholars was much more evident. In Babylon and Uruk, for example, students speaking Aramaic as their mother tongue, and fluent in Greek, had to master Sumerian and Akkadian to access textbooks in mathematics, astronomy, divination, and medicine. They were born in highly educated families, in which knowledge was passed down from generation to generation. The mathematics they developed was by then closely associated with astronomy. In Egypt as well, scholars assimilated foreign knowledge and developed their mathematics in new directions.

CHAPTER THREE

Inventing Mathematics

The Creative Work of Scribes and Authors

CHRISTINE PROUST AND REVIEL NETZ

It is not always easy to decide whether a particular text contains a specific invention. We do not know exactly when notions or methods appearing at a given time in a given document were invented or even which results or practices were perceived as innovative by their authors. Indeed, it is not always clear whether authors thought of themselves as inventors, that is, as true authors. This chapter presents mathematical texts that show evidence of novelties or breaks from those attested in previous periods, at least in our eyes as modern historians. In doing so, culture in which those novelties and breaks occurred is also brought into relief. We shall begin with Mesopotamia and then move on to Greece.[1]

THE SEXAGESIMAL PLACE-VALUE SYSTEM IN DIFFERENT CONTEXTS IN MESOPOTAMIA

Most of the mathematical cuneiform texts were produced in southern Iraq, but very active mathematical centers have also produced small *corpora* in the upper Tigris and Euphrates valleys, in Syria, and in western Iran. The mathematical landscape that can be perceived from the available sources is a succession of scenes, sometimes richly documented, sometimes rather vague. In some instances, transmissions, continuities, or conceptual shifts between moments or places of invention can be described. But before we travel through various contexts in which these changes occurred, let us review the principles of the sexagesimal place-value system and notation which are, in one form or another, at the heart of mathematical concepts and practices.

Mathematical cuneiform texts that date from the end of the third millennium to the end of the first millennium used a positional base 60 numerical system (the so-called sexagesimal place-value notation, or SPVN) consisting of two signs: a wedge (𒁹) for a unit, and a chevron (𒌋) for ten units. In a number noted in SPVN, each wedge represents sixty times more than the wedge in the previous position (at the right), and each chevron represents sixty times more than the chevron in the previous position. The system is well illustrated by the multiplication table by 9 in the cuneiform tablet HS 217 (Figure 3.1).

In line 7 of the obverse, we see that the result of 9 times 7 is noted 𒁹 𒐈 (1.3); in this two-place number, the wedge in the left place represents sixty times each wedge in the right place. In line 6 of the reverse, we see that the result of 9 times 20 is noted 𒐈 (3). Thus, 3×60 is noted in the same way as 3 (line 3 of the obverse). We see that the cuneiform notation does not indicate the place of the units in the numbers; in other words, there is no sign equivalent to the digit "0" at the beginning or the end of a number. The notation is "floating." As we shall see, the absence of a specific sign indicating empty places at the beginning or end of a number was not a problem in the context of the culture

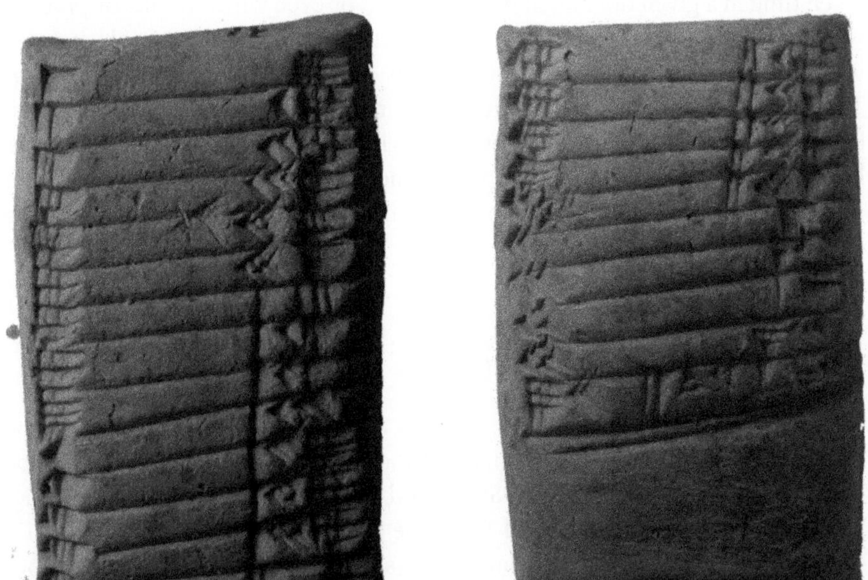

FIGURE 3.1. HS 217. Photo by Christine Proust courtesy of the Hilprecht Collection, University of Jena, Germany.

of calculation prevailing in the scribal schools. However, the lack of indication of empty places *within* a number was a serious problem, causing ambiguities and errors. This problem was solved already in the Old Babylonian period by using spaces between digits, and later by using a specific sign (see below).

The sexagesimal system used in these texts is similar to that we use today for counting time but presents some differences. In the ancient system, we have the following:

- The number of positions is not limited (in some Late Babylonian texts, numbers exceeding twenty positions are attested).
- The position of the units is not indicated (floating numbers).
- Numbers are used primarily for multiplication.

The expression "sexagesimal place-value notation" (SPVN) integrates their three fundamental features: base 60, place-valued principle, and floating notation. As a matter of convention, the sexagesimal digits are represented today by modern Indo-Arabic figures; the places are separated by dots. For example, we write the results of 7×9 and 50×9 in the multiplication table (Figure 3.1) as 1.3 and 7.30, respectively.

INVENTING MATHEMATICS TO CONTROL THE LANDS: EVALUATING SURFACES IN THE THIRD MILLENNIUM

Mesopotamia of the third millennium underwent important political transformations with the rise of powerful cities, such as Uruk, and subsequently imperial dynasties controlling vast territories, such as the Sargon Dynasty and the Third Dynasty of Ur (see Figure A1 Map of Mesopotamia in the Appendix).

Groups of scholars at the service of the royal courts undertook major reforms affecting writing, law, accounting, and metrology.[2] It is in this context that the sexagesimal place-value notation was developed, the arithmetic advantages of which would be exploited by the scholars of the following generations to produce highly sophisticated mathematical elaborations. What exactly were the mathematical problems that led to the invention of the sexagesimal place-value notation? A handful of mathematical tablets show that these problems deal essentially with the evaluation of surfaces.

Among the earliest known mathematical texts are five tablets providing the measurements of the sides of squares or rectangles and the corresponding surface measurements. These five tablets come from a small geographical area situated inside the triangle of Nippur, Lagash, and Uruk, in southern Mesopotamia, and are dated to a relatively short period between 2600 and 2300 BCE.[3]

These tablets form the main part of the mathematical *corpus* known for the Early Dynastic (ED) periods. Their common point is that they relate lengths to surfaces, and that they date to the Early Dynastic periods, that is, *long before the invention of the sexagesimal place-value notation*. Considerable research has been devoted to one or several of these texts, notably by Jöran Friberg. Without going into details, let us just give some elements of description of these tables that shed light on the nature of the mathematical problems that arose at that time in evaluating surfaces. The earliest texts (dated ED IIIa) give the surfaces of large square or rectangular territories, whose sides most often measure several kilometers, while the most recent (ED IIIb) give the surfaces of small squares or rectangles, whose sides measure a few centimeters. These differences in magnitude reveal changes in the way surfaces were measured, in the metrology shaped for these measurements and calculations, and in the concept of space: from a global assessment of very large lands, the notion of surface tended to be generalized to all kinds of two-dimensional extensions, whether large or small, agrarian or geometric. More importantly, the evaluation of small surfaces raises a particular problem: the scale of the units of surface does not match with that of the units of length. Only one unit of surface is derived from units of length: 1 *sar* is the measurement of a square of 1 *ninda*-side (approx. 6 m). Consequently, evaluating the surfaces of squares or rectangles with sides less than 1 *ninda* rely on fractional calculation.

To solve this difficulty, the authors of tablet CUNES 50-8-1 used successive sexagesimal subdivisions of the unit *sar*. This is the oldest attestation of the use of sexagesimal fractions in order to calculate surfaces. At the same time, sexagesimal multiples of the unit *ninda* were used to express long distances. Such a sexagesimal scale had already been used widely since the birth of writing, and probably before, to count things, people, or animals. This non-positional sexagesimal system, of additive principle, has been called "System S":

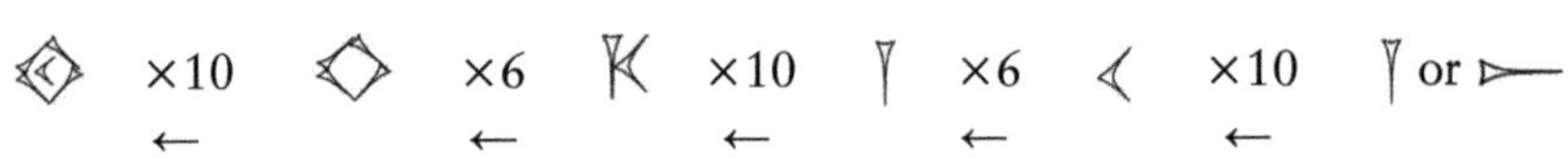

šar'u	šar$_2$	geš'u	geš$_2$	u	diš or aš
36,000	3,600	600	60	10	1

System S (sexagesimal additive)

The tablet shows that the factor sixty used to produce sexagesimal multiples of the units was extended to divide by sixty, producing sexagesimal sub-multiples of the unit *sar*. The calculation needed the knowledge of equivalences between usual fractions (1/2, 1/3, 1/4, etc.) and sexagesimal fractions (1/60,

$1/60^2$, $1/60^3$). This scale of sexagesimal fractions of *sar*, namely a sixtieth of *sar*, sixtieth of sixtieth of *sar*, and sixtieth of sixtieth of sixtieth of *sar*, do not seem to have spread widely (to date, only tablet CUNES 50-8-1 testifies to these three levels of subdivisions by sixty). However, the major innovation, which is the subdivision of the unit into sixtieths, giving a unit called "*gin*," was developed in the following periods. For example, ½ unit is 30 *gin*, 1/3 unit is 20 *gin*, 1/5 is 12 *gin*, and so on (see below).

Documents dated to the Old Akkadian period shed light on new practices. These are small tablets presenting a class of mathematical problems in which the measurements of the surface and the length of a rectangular field are given, and the width must be found. For example, Nippur's HS 818 tablet runs as follows:

> 67 ½ *ninda* the length
> 1 *gan* (the surface). (What is) the width?
> The width is 1 *ninda* 5 *kuš* 2 *šu-du-a* 3 1/3 *šu-si*[4]

The surprising point here is that the length, i.e. the value by which the surface must be divided to obtain the width, corresponds to the number 1.7.30 in sexagesimal place-value notation (see Table 3.1), and is a regular number in base 60 (the prime factors are powers of 2, 3, or 5; see below). This cannot be a coincidence, even less so when we observe that in all exercises of the same type of the same period, the length also corresponds to a regular number in sexagesimal place-value notation.

The fact that the divisor (corresponding to the length by which the area is to be divided) can be represented by a regular number in base 60 undoubtedly reflects particular calculation practices. This feature suggests that the calculation was based on the base 60, and that a technique of division as multiplication by the reciprocal was already in use. In this case, it can be inferred that a kind of sexagesimal place-value system was already in use in the Old Akkadian period. However, the sexagesimal place-value notation is not attested in the sources until the Ur III period, at least one century later. This apparent paradox, as well

TABLE 3.1. Old Akkadian field texts: The lengths of the fields

Museum or publi. No[5]	*Provenience*	*Length*	*Corresponding SPVN*	*Primary factors: Powers of*
ZA 094, 003	Umma	2 *ninda*	2	2
CMAA 16-C5	Unknown	9×60 *ninda*	9	3
PUL 29	Girsu	2×60+40 *ninda*	2.40	2, 5
PUL 31	Girsu	4×60+3 *ninda*	4.3	3
HS 815	Nippur	60+7 ½ *ninda*	1.7.30	2, 3, 5

as other paleographic, lexical, or mathematical evidence, have led to the conclusion that some sexagesimal place-valued techniques of calculation were practiced on material instruments, such as token, sand drawings, or abacus, since the Old Akkadian period.

The oldest known texts where the sexagesimal place-value notation is used systematically are reciprocal tables from southern Mesopotamia dated to the Ur III period, for example, the Nippur HS 201 tablet (see Table 3.2).

TABLE 3.2. Tablet HS 201 containing a table of reciprocals

Obverse

The reciprocal of 2 is 30		
[3]	reciprocal	20
[4]	reciprocal	15
[5]	reciprocal	12
[6]	[reciprocal]	10
[7]	[reciprocal]	no
[8]	reciprocal	[7].30
10 minus l	reciprocal	6.40
10	reciprocal	6
11	reciprocal	no
12	reciprocal	5
13	reciprocal	no
14	reciprocal	no
15	reciprocal	4
16	reciprocal	3.45

Reverse

20 minus 3	reciprocal	no
20 minus 2	reciprocal	3.20
20 minus 1	reciprocal	no
20	reciprocal	3
21	reciprocal	no
22	reciprocal	no
23	reciprocal	no
24	reciprocal	2.30
25	reciprocal	2.24
26	reciprocal	no
30 minus 3	reciprocal	2.13.20
30 minus 2	reciprocal	no
30 minus 1	reciprocal	no
30	reciprocal	2
31	reciprocal	no
32	reciprocal	1.52.30

The Ur III reciprocal tables make apparent the fundamental distinction between two categories of numbers, the regular numbers which admit a reciprocal (*igi*) with a finite number of positions in base 60 (their prime factors are powers of 2, 3, or 5), and the non-regular numbers, which do not admit a finite reciprocal ("*igi nu*" in Sumerian, that is, "no reciprocal").

In conclusion, the sexagesimal place-value system would have existed in material form as a calculation device long before it appeared in the texts in the written form. The examples quoted above show the complexity of the history of the sexagesimal place-value notation, which results from several different innovations: the construction of sexagesimal multiples of numbers, the construction of sexagesimal sub-multiples of measurement units (for example, the unit *gin*), the implementation of non-positional sexagesimal calculation techniques (for example, in tablet CUNES 50-8-1), the use of a material device (for example, for solving Old Akkadian field problems), and finally the development of the sexagesimal place-value notation proper.

INVENTING MATHEMATICS IN THE OLD BABYLONIAN PERIOD: SCRIBAL SCHOOLS AS SCHOLARLY CENTERS

The major intellectual phenomenon that marked the early second millennium is the explosion of scribal schools in the ancient Near East. The vast majority of the cuneiform mathematical texts that have come down to us were produced in the context of these schools. Around 2,000 tablets inform us about mathematical practices in schools or in relation to them. This documentation consists mainly of school exercises, but also of scholarly texts revealing advanced mathematical methods that, using anachronistic terms, we recognize today as first- and second-degree equations, theorem of Pythagoras, theorems of proportionality, number theory, and others. It is important to emphasize that although we use terms such as "school" and "student," the knowledge implied by these mathematical texts should not be thought of as the material of general numeracy: as discussed in Chapter 2, these were students being groomed for a very elite profession. What is meant here by "school texts" are texts written by or for young students which reflect the early stages of education, the so-called "elementary" and "intermediate" levels.

The mathematical education of elementary and intermediate levels focused on two essential tools: first, the role of the SPVN as a tool for multiplications and divisions; second, the role of metrological tables as a tool to solve mathematical problems dealing with quantities (see Chapter 2 for details).

Advanced mathematical problems deal with a variety of topics such as field sizes, brick transport, digging of canals, grain storage, building walls, buying and selling, profits and interest, and many others. Despite this realistic appearance, however, cuneiform mathematics cannot be considered as purely practical. Indeed, the situations described in many problems are unrealistic, and the vocabulary and syntax used to solve the problems belong to an artificial technical language, elaborated only for mathematics. Even when the statements of problems refer to concrete situations, the situation described is often quite artificial. Moreover, an important part of the mathematical *corpus* has nothing to do with *real life*. Some of the mathematical texts are based on numerical algorithms that exploit the specific properties of base 60, rich in divisors. The most common of these algorithms is based on factorization processes, in order to calculate reciprocals, square roots, or cubic roots. An example of abstract mathematics is a list of Pythagorean triples in tablet Plimpton 322 (Figure 3.2).

The masters of the scribal schools appear to be inventive scholars, little concerned with everyday vicissitudes, but good connoisseurs of the affairs of the city. Strong relationships must have existed between scholarly circles and administrators, officials, governors, or members of the royal court. Some scribal schools, for example in Nippur, appear to have been not only places of teaching, but also scholarly centers, crucibles of the intellectual life of the city.

FIGURE 3.2. Plimpton 322: "Pythagorean Triples." Photo by agefotostock / Alamy Stock Photo.

ALGEBRA IN CUNEIFORM SOURCES?

The question of the algebraic nature of cuneiform mathematics and its historiographic roots was discussed in depth by Jens Høyrup. This question could be reduced to the following: are the procedures for solving the quadratic problems found in cuneiform sources based on a mathematical reasoning, and if so, what is the nature of this reasoning? This question is at the heart of Jens Høyrup's works, particularly his seminal book *Lengths, Widths, Surfaces* (2002), the main arguments of which are briefly summarized below.

It is often considered that the ancient scribes discovered the sequence of operations necessary to solve the quadratic problems by trial and error or by chance, rather than by reasoning. However, Otto Neugebauer and other historians after him showed the systematic and general nature of the procedures used in mathematical cuneiform texts, despite the fact that these procedures are exposed in these texts through numerical examples assuming a paradigmatic function. To justify the procedures in their own eyes, historians transformed the sequence of operations prescribed in cuneiform texts into formulas using the symbolism of modern algebra. In his book, Høyrup proposes a new approach, which consists of reading as closely as possible the cuneiform texts in order to carry out a meticulous linguistic analysis, making it possible to identify a genuine technical vocabulary and to elucidate the meaning of each of the terms. Høyrup observes, in particular, that the scribes used several different expressions to describe what we believe to be the same operation. However, he shows that the different terms refer to different concepts. For example, a multiplication may be a simple arithmetic operation (Sumerian ideograms a-ra$_2$, which can be translated "times"), but it may also refer to an operation of a geometric nature, which consists in crossing a length and a width to construct a rectangle (Akkadian verb *šutakûlum*). Applying this "close reading" to procedure texts, he gives rise to a completely new interpretation: the scribes followed a figurative reasoning, and the steps of the procedure can be interpreted as a succession of "cut-and-paste" operations on squares, rectangles, and gnomons. Babylonian "algebra" is, in fact, according to Høyrup, a "naive geometry" which foreshadows methods developed long afterwards by Greek and Arab mathematicians for solving quadratic problems.

To illustrate his approach, Høyrup often gives as an example the first problem of a tablet kept at the British Museum, BM 13901 (CDLI number P254406). This tablet is of great importance in the historiography as it contains 24 problems organized systematically in order to review all facets of solving quadratic problems. Caveing (1994: 35, 83) considered this text as a kind of "textbook of calculation of the second degree" and compared it with Book II of the Elements of Euclid.

The following is a translation of BM 13901 #1 by Høyrup (2002: 50):

1. The surface and my confrontation I have accumulated: 45' is it. 1, the projection,
2. you posit. The moiety of 1 you break, 30' and 30' you make hold.
3. 15' to 45' you append: by 1, 1 is equalside. 30' which you have made hold
4. in the inside of 1 you tear out: 30' the confrontation.

The representations of the problem and its solution by Thureau-Dangin, Neugebauer, and Høyrup makes clear the differences of interpretation between the former two, who adopt an algebraic approach, and Høyrup, who adopts a geometrical approach (see Diagram 3.1).

Høyrup's interpretation, which gives a geometric meaning to the operations prescribed in the procedures, is now widely accepted by specialists. His approach has profoundly transformed our vision of cuneiform mathematics.

Thureau-Dangin (1938: xxi) (x is the side to be sought)

$$x^2 + x = 45'$$

$$x = \sqrt{(1/2)^2 + 45'} - 1/2 = 30'$$

Neugebauer (1935–7: vol. 3, 10)

$$\text{Equation: } x + x^2 = a$$

$$\text{Solution: } X = x = \sqrt{\left(\frac{1}{2}\right)^2 + a} - \frac{1}{2}$$

Høyrup (2002: 51) (s is the side to be sought)

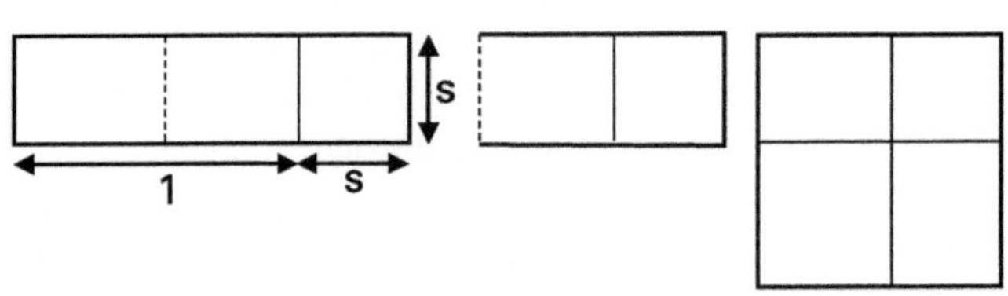

DIAGRAM 3.1. Three approaches of BM 3901 #1.

INDETERMINATE ANALYSIS IN CUNEIFORM SOURCES?

The clay tablet registered under number 322 in the Plimpton Collection (CDLI number P254790), now housed at Columbia University, New York, is, for a wide audience, the star of the mathematical cuneiform sources. Since its publication by Neugebauer and Sachs in 1945, it has been the subject of many specialized articles and continues to fascinate mathematicians and computer scientists. This fascination undoubtedly comes from the fact that this text can be interpreted as reflecting a systematic search for the so-called "Pythagorean triples" (that is, three whole numbers x, y, z such that $x^2+y^2=z^2$), and therefore as falling within the scope of number theory. Can the Plimpton 322 tablet be considered as reflecting remarkable discoveries in indeterminate analysis or, more generally, in number theory? The answers given by the scholars who studied this text are divergent. For example, Neugebauer and Sachs (1945: 40) saw in Plimpton 322 evidence that "Babylonian number theory was acquainted with rules (which) produce Pythagorean numbers, i.e., a theorem like Euclid X 29 lemma 1," while Robson (2001) cooled such enthusiasm, and Britton, Proust, and Shnider (2011) revived it. The following interpretation adopts this latter point of view.

The tablet was unearthed in Iraq in the course of illegal excavations at the beginning of the twentieth century. It was subsequently bought by the adventurer Edgar Banks (1866–1945), a colorful figure who was the inspiration for Indiana Jones. Banks sold the tablet in 1922 or 1923 to American collector George Plimpton. The origin of the tablet is uncertain. Banks claimed the tablet came from Larsa. Its paleography and orthography features suggest that the tablet date to the Old Babylonian period and originates from southern Mesopotamia. When it entered the Plimpton collection, the tablet was registered as administrative. It was Neugebauer who discovered its mathematical character.

The tablet is displayed in the unusual landscape format and contains on the obverse four columns of numbers separated by vertical lines; the reverse contains only the continuation of the vertical lines. The profiles of the upper and lower edges show that the left side of the tablet, about one-third of the original document, is lost. The limited space of this chapter does not allow to give a detailed description of the table, which is easy to find in many publications. Looking at the first column is sufficient to shed light on the nature of the document.

The heading of the first column indicates its content as follows:

> The square of the diagonal from which 1 is subtracted, so that the width comes up

The first column contains a list of 15 numbers in SPVN. For example, in the fifth row, the entry is the number 1.48.54.1.40. This number is the square of 1.20.50.[6] Moreover, if 1 (the first digit to the left) is subtracted, it comes out as 48.54.1.40, the square of 54.10. We see that the fifth entry conforms exactly to the description of the heading, and it is the same for the other entries: each number in the first column is the square of a number (the diagonal), such that if 1 (the length) is subtracted, the number obtained is again the square of a number (the width), providing 15 sexagesimal unitary rectangles (rectangle which length, width and diagonal are finite sexagesimal numbers, and the length is 1). The fact that the length is always 1 guarantees that the 15 corresponding Pythagorean triples are all independent of each other.

How were the numbers of the first column produced which bear such miraculous properties? This question was controversial until Friberg (2007: appendix 8) published a mathematical tablet of the Schøyen collection closely related to Plimpton 322. Tablet MS 3971, dated to the Old Babylonian period and from unknown provenance (probably southern Mesopotamia), is particularly illuminating.[7] Problem 3 of this tablet explains in detail the way in which a "Pythagorean triple" (or, less anachronistically, a sexagesimal unitary rectangle) can be obtained from a pair of reciprocals.[8]

The question now is how the reciprocal pairs that produced the "Pythagorean triples" were selected: were they chosen at random? Are they the result of a systematic selection process? Is this list complete (according to criteria to be defined)? The computer scientist and historian of sciences, Derek J. de Solla Price, observing that the vertical lines separating the columns on the obverse continue on the reverse, concluded that the 15 rows inscribed on the obverse of Plimpton 322 were probably the first ones of a larger table, which was intended to cover also the lower edge and the reverse of the tablet. Price identified a list of 38 values of "parameters" (pairs of integer numbers he labeled *r* and *s*) from which the Plimpton 322 table could be constructed. The first 15 of them generate precisely the 15 existing rows. The remaining 23 pairs could generate 23 other rows, which could take up just the space available on the lower edge and reverse of the tablet. These "parameters" and subsequent calculations and reasoning were developed by Price with modern methods and concepts (approximations, inequalities, and fractions). However, a list of 38 pairs of reciprocals (corresponding to the quotients *r*/*s* and *s*/*r* of Price's parameters) could be produced by simple operations on lists of numbers extracted from the reciprocal table.

It is thus possible to produce the exact list of pairs of reciprocals on which Plimpton 322 is based using only methods and concepts familiar to the scholars of the ancient scribal schools. This possibility provides a powerful argument in favor of the hypothesis that the Plimpton 322 tablet was the result of a systematic search for the complete list of solutions to an undetermined problem. In addition, it offers the possibility of rebuilding the end of the unfinished table, as well as the lost left-side part.

These observations and speculations suggest that the entries of Plimpton 322 are not a list of data randomly collected, and stored for uses such as teaching, but a problem text. This would be an undetermined problem, consisting of finding all the sexagesimal unitary rectangles described in the heading of the first column, and an attempt (unfinished) to provide a complete solution in tabular format.

INVENTING A NEW LANGUAGE OF MATHEMATICS

At the end of the Old Babylonian period, under the last rulers of the Hammurabi Dynasty, the southern Mesopotamian cities revolted against the power of Babylon, resulting in wars and destruction, and finally the exodus of populations. Groups of scholars migrated from southern Mesopotamia to the cities of Kiš and Sippar in central Mesopotamia. Some of these groups continued a scholarly activity, including mathematics. Their mathematical production shows that the communities of scholars resettled at Kiš or Sippar had preserved the knowledge and traditions developed in southern Mesopotamia before the revolts, but these texts also reflect new mathematical explorations. A new way of writing mathematics, which one could call "quasi-algebraic," has emerged in this context.

The mathematical texts dated to this time are long lists of several hundred statements of problems, each list covering dozens of numbered tablets like the pages of a book. This practice of writing extremely long texts on numbered tablets, the so-called "series" to use the vocabulary of Assyriologists, was usual in the first millennium, but rare in the Old Babylonian period. Twenty mathematical series texts are known to date. Museographical, paleographical and textual evidence show that these tablets probably come from the same region or even the same city located in central Mesopotamia.

One of the remarkable aspects of this mathematical *corpus* is its writing system. The language spoken in Mesopotamia during the Old Babylonian period was Akkadian, a Semitic language. However, at that time, the scholarly and administrative texts used cuneiform signs representing the words of the Sumerian language, sumerograms, often inserted in Akkadian texts. Sumerian is a language of an unknown linguistic group, which was spoken in Mesopotamia in the third millennium, and which then disappeared as a living language while remaining until the abandonment of cuneiform writing at the beginning of our era as a written language for scholarship. Mathematical series are written almost exclusively with sumerograms in an extremely concise style. Neugebauer claimed that the writing of series texts was a pure mathematical symbolism unrelated to a spoken language. Thureau-Dangin thought that these texts were to be read in Akkadian. This would be an example of what is named today as an "alloglottographic" writing, that is, a writing which does not represent the language in which the text is pronounced. Other historians consider that series texts were really written in Sumerian.

The uniqueness of the style concerns not only the language, but also the architecture of the texts. To have an idea of such structure, let us look at an extract of one of the mathematical series texts kept in the Louvre, AO 9071 (CDLI number P254392), which is the seventh tablet of a series according to its colophon.

The 59th statement of this tablet is simply, "I subtracted, it's 45." This statement, which at first glance seems somewhat incomplete, can only be understood in the context of a large tree-like textual structure. Its principle is as follows: when a given information is common to several successive statements, this information is provided in the first statement of the group, but not repeated in the following ones. This principle is applied at different levels, so much so that the further we go in the text, the more information is locally incomplete. The elements of an effective statement are thus dispersed throughout the vast list of statements, but in a perfectly planned manner, allowing the expert reader to find them. For example, when all the information related to the statement quoted above ("I subtracted, it's 45") has been retrieved, this statement appears to refer to a highly complex problem, which could be expressed as follows in modern algebraic language (x represents the length of a rectangle, and y its width).

$$\begin{cases} (3x+2y)\dfrac{1}{13}+x=40 \\ -\left\{\begin{array}{l} \left\langle\left\{[(x+25)+(y+1.30-x)+(x+y+35)]\dfrac{1}{11}+4x\right\}\dfrac{1}{7}+y\right\rangle\times \\ 2\times\dfrac{1}{16}-(x+2y)+x+y+(3x-2y) \end{array}\right\}\dfrac{1}{7}+(x+y)=45 \end{cases}$$

The structure, syntax, and terminology of the text, although originally inspired by those of the Sumerian language, are artificial linguistic creations designed to express algebraic relationships as complex as desired. In particular, this syntax allowed the ancient scribes to represent hierarchies of operations that, in modern algebraic language, we represent by levels of parentheses, here four levels.

The combinatorial processes that generate the series texts permits an almost unlimited development of lists of statements. It seems that their authors were particularly interested in the potentially of unlimited extension of the process they had invented. Indeed, we know higher-order series which bring together several series that Neugebauer (1935–7: vol. 1, 385) called "super series," counting thousands of statements. The conciseness of the writing made it possible to concentrate hundreds of statements in the limited space of relatively small clay tablets. For example, tablet A 24194 measures a few centimeters on each side and contains more than 200 statements.

This astonishing virtuosity in the art of developing tree-structured lists seems to have been cultivated in the learned circles of immigrants from southern to central Mesopotamia at the end of the Old Babylonian period. These scholars were well acquainted with the mathematical traditions of the south which had been developed in the context of scribal schools and reproduced the terminology and themes of these traditions. At the same time, the series are, in some respects, totally innovative.

Other contemporary genres of texts exhibit similar features, in particular serial organization and tree structures. It is the case of divination texts, perhaps written by the same community of scholars. These writing techniques may reflect the emergence of a new practice of preserving and archiving scholarly texts, the libraries. A new scribal culture was emerging, but it remained latent until a later period.

NEW MATHEMATICAL CULTURES IN LATE PERIODS

The Late Babylonian period, which includes the Achaemenid and Hellenistic eras, begins as we move into the second half of the first millennium. After a long period in which few mathematical tablets have come down to us, during this later period we begin to see again mathematical tablets, about 100 of them. They come mainly from the scholarly libraries and archives discovered at Uruk and Babylon, as well as, to a lesser extent, from cities such as Kiš, Sippar, or Nippur; very few come from Assyria. These tablets contain a variety of texts, including metrological and numerical tables, coefficient lists, and lists of problems with detailed procedures. Elaboration and transmission of mathematics, along with astral sciences, medicine, and other scholarly domains, were the appanage of high priests.

The tablets found in Uruk and dated to the Achaemenid period were written, copied, or owned by representants of the most prominent family of exorcist-physicians (*āšipu*) of the town. The tablets show how these priests revived the ancient metrological knowledge dated back to the Old Babylonian period and, at the same time, elaborated new technical tools for quantifying, buying, and selling land.

The Hellenistic period is marked by the development of mathematical astronomy. Mathematical texts produced in Uruk and Babylon during this time are quite different from those produced in the previous period. Mathematical texts do not exhibit connections with practical arithmetic and metrology in use in Late Babylonian cities, like in the Achaemenid period, but rather they reflect a revival of the Old Babylonian traditions. Numerical tables show a strong interest in the numerical calculation in sexagesimal place-value notation. For example, large tables of reciprocals of many-places sexagesimal numbers were elaborated. This section focuses on one of these tables of reciprocals.

AO 6456 (CDLI number P254386), on display at the Louvre Museum, is one of the best known and most spectacular tablets among mathematical cuneiform documents. The tablet contains a table giving numbers starting with the digits 1 and 2 and their reciprocals. The numbers are of very large size, some of them exceed ten sexagesimal positions. Empty sexagesimal positions inside a number are indicated by a special sign, , newly introduced in the cuneiform writing, which can be interpreted as the digit "zero." It was used also as a separation sign.

Interesting information on the context of writing of this tablet is provided by marginal inscriptions. An invocation to the gods Anu and Antu noted on the upper edge reflects the religious context. A colophon noted at the end of the text runs as follows:

> The reciprocal of 3 is 20
>
> First section: 1 // *amû* // 2 // *amû*.[9] Unfinished.
>
> Tablet of Nidinti-Anu descendent of Ina-qibit-Anu, descendent of Hunzû, incantator of Anu and Amtu, Urukean. Hand of Qibit-Anu, his son.

The colophon makes clear the nature of the table and provides information on its authors. The enigmatic "1 // *amû* // 2 // *amû*" can be interpreted as meaning "list of reciprocals of numbers beginning with 1 or 2." The mention "First section" indicates that AO 6456 is probably the first tablet in a series of several successive tablets. This explains the mention, "Not finished," which suggests that other tablets are supposed to follow, so we expect the reciprocal tables of numbers beginning with 3, 4, etc. This series context gives the meaning of the first line of the colophon, "the reciprocal of 3 is 20," which quotes the first line of the following tablet, and thus serves as a catchline.

The colophon indicates the name of the "owner" of the tablet, Nidinti-Anu, his genealogy and profession, "incantatory priest of Anu and Amtu," that is, a high priest belonging to the clergy of the Bel Temple in Uruk. The "owner" of a tablet was in fact the supervisor of the scribe who had copied it. This tablet shed light on the process of training learned priests of the temple of Bel. It is in this environment that the use of cuneiform writing was perpetuated, a practice that was already disappearing in other segments of society in favor of the alphabetical transcription of Aramaic language, using ink on skin or papyrus.

These colophons then provide a window of the cuneiform culture in which the idea of a mathematical author has an identity, but in a circumscribed fashion (see also the discussion of colophons in Chapter 2). As we move to Greek and Roman mathematics, a very different picture emerges.

THE GREEK INVENTION OF THE AUTHOR

At some point in the late fifth century BCE, Hippocrates of Chios (not to be confused with the father of medicine, Hippocrates of Cos) began to circulate a papyrus with some mathematical results. It appears that those dealt with "lunules"—the areas intercepted between two arcs of different circles (both concave in the same direction). We may concentrate on the first and simplest example. An isosceles, right-angled triangle is inscribed in a semicircle, so that two equal arcs are intercepted by the two smaller sides. A third arc is drawn on the hypotenuse, similar in shape to the two smaller ones. A lunule is thus intercepted between the two arcs in Diagram 3.2.

There are three areas intercepted between an arc and a line (two on top, one at the bottom) and since we know that the three are of similar shapes, we can also see that the three areas are to each other as the squares on their lines. Since the two squares on the lines at the top are equal to the square on the line at the bottom (by the so-called Pythagoras' theorem), we know that the two areas at the top are equal to the area at the bottom. Now, if we take the triangle, remove from it the area at the bottom, and add to it the two areas at the top, it becomes the lunule. Since the area at the bottom is equal to the two areas at the top, we find that the triangle is equal to the lunule: a circle-like figure, found equal to one bounded by straight lines.

This is the earliest evidence we have for Greek mathematics (the references to mathematical theorems by Thales and Pythagoras are later and probably spurious; although, of course, craft and educational knowledge of

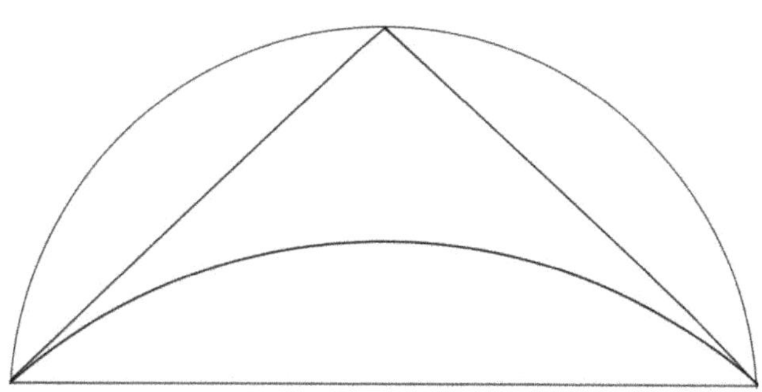

DIAGRAM 3.2. Simple case of a lunule.

mathematics—mostly ignored in this brief survey—certainly circulated for generations before). Several observations stand out. First, the mathematics is indeed elementary: one must know Pythagoras' theorem (but this was probably common knowledge in the Mediterranean world and is not very far from results concerning rectangles already familiar in Mesopotamia, more than a millennium before). The key idea of the proof—that similar areas are to each other as the squares on any of their lines—is sophisticated but essentially intuitive. (Think of each circular area as a similar "bite" from its respective square, and the idea is indeed obvious.) Second, and the converse to the first: this proposition only makes sense relative to a fairly well-established basis of numeracy and familiarity with geometrical shapes—which is to be understood as the legacy of the ancient Near East. Third, the argument is striking: the result is extremely surprising (coming close, as it does, to the squaring of the circle—which must be well known for its difficulty for anyone trained in measuring geometrical shapes so, once again, assuming some basic education). But it is also a rigorous, compelling proof—coming as it does from a civilization of debate and persuasion—reflective, of course, of the practices of the Greek city state. It is natural that early publications of mathematics will include impressive results—well argued, surprising claims. Why would anyone circulate boring mathematics? Fourth, and finally, this is not merely circulated but, specifically, this is circulated under the name of Hippocrates of Chios. This author sought fame. His work was sufficiently original and impressive to gain attention and to be remembered as due to that particular author; he would be referred to, by name, by future authors and so, ultimately, memory of his achievement did survive (though not his original work).

This is the most fundamental invention of Greek mathematics: the invention of the mathematical author. This is, indeed, part of a more general cultural pattern. The universal tendency in early civilizations is for literature to circulate anonymously or pseudonymously. Canonical texts, whether written or orally transmitted, are understood to represent timeless truths and so should be better endowed with the quasi-divine, prophetic origins. "Sing, Goddess, the rage of Achilles . . ."—thus began the Iliad, the most important poem regularly recited by the Greeks. The Israelites had their prophets and anonymously authored histories; the Mesopotamians, too, had their epics and instructional canonical texts, all understood to come from divine sources. The Vedas—the earliest origin of some Indian mathematical sutras—were obviously divine. Even more down-to-earth works such as the Chinese *Nine Chapters on the Mathematical Art* circulated without a named author. Otherwise, how could they be canonical? In most ancient cultures, the canon was unauthored and named authorship was associated more with commentary to the canon. Early on, a new Greek model emerged—associated, once again, with the more egalitarian and competitive experience of the Greek city state—

of sending out one's works with a strong, personal voice attached to them. The named author emerges in lyric poetry in the seventh century BCE, it defines a new kind of wisdom literature—what ultimately came to be "philosophy"—through the sixth and fifth centuries BCE. In the later fifth century BCE, at a moment of special florescence of Greek civilization, the model of authored literature became widespread. Greeks must have discussed and argued orally concerning numbers and geometrical figures for many decades before, especially around their practice of elementary education. In the later fifth century BCE, a new genre emerged—writing on mathematics, in particular on geometry, with a strong authorial voice, typically emphasizing one's original contributions. The word "invention" begins to be directly applicable: writing in one's own voice, seeking fame for one's distinctive contribution, it is only natural that the new type of texts—proudly authored mathematics—will aim to find something new. The lunule might have been the first; it was certainly not the last.

THE INVENTION OF THE GENRE

While Greek mathematicians wrote in their own names—trying to put their own individual seal—they did so within a fixed literary mold. This is typical of Greek literature as a whole. The dramatists competed against each other—in tragedies that ultimately followed the exact same rules (three actors and a chorus, dialogue in a fixed meter, choral interludes . . .). This is not just an analogy but also an important context for the understanding of ancient science. The habits of Greek literature emerged out of rule-based performances and for this reason the idea took hold, early on, that a literary practice ought to be bound by the rules of a genre. As people started to put mathematics into writing—to make it into literature—it was only natural that a mathematical genre would emerge, as well. To see this, let us consider an extended piece of Greek mathematics, and I cite a brief proposition from Euclid's *Elements* (I.35; Diagram 3.3).

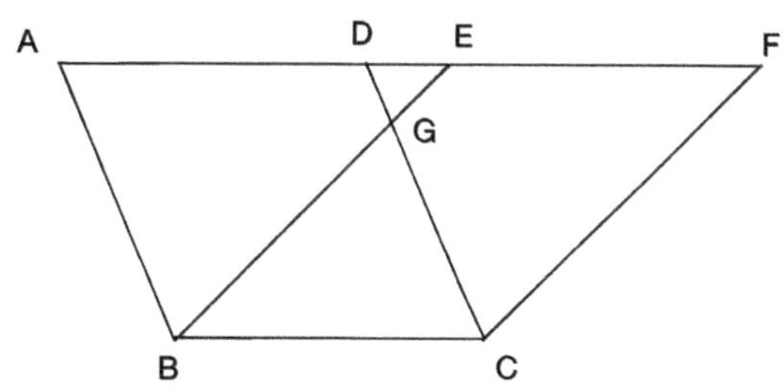

DIAGRAM 3.3. Euclid's *Elements*, Book I, Proposition 35.

> Parallelograms which are on the same base and in the same parallels are equal to one another.
>
> Let ABCD, EBCF be parallelograms on the same base BC and in the same parallels AF, BC;
>
> I say that ABCD is equal to the parallelogram EBCF.
>
> For, since ABCD is a parallelogram, AD is equal to BC. For the same reason also EF is equal to BC, so that AD is also equal to EF.
>
> And DE is common; therefore the whole AE is equal to the whole DF.
>
> But AB is also equal to DC; therefore the two sides EA, AB are equal to the two sides FD, DC respectively, and the angle FDC is equal to the angle EAB, the exterior to the interior; therefore the base EB is equal to the base FC, and the triangle EAB will be equal to the triangle FDC.
>
> Let DGE be subtracted from each;
>
> Therefore the trapezium ABGD which remains is equal to the trapezium EGCF which remains.
>
> Let the triangle GBC be added to each;
>
> Therefore the whole parallelogram ABCD is equal to the whole parallelogram EBCF.
>
> Therefore parallelograms which are on the same base and in the same parallels are equal to one another; which it was required to prove.

We should note the repetitive structures, such as "let X be subtracted/added from/to each"; and fixed phrases, such as "the X which remains" or "the whole X." The language of Greek mathematics is sparse and repetitive, made not so much of individual words as of entire formulaic expressions. Crucially, while this is a single example, the very basic linguistic pattern can be found repeated anywhere else in the extant Greek mathematical corpus. Even expressions which occur only once in this proposition—such as, "I say that . . ." are, in the corpus as a whole, extremely common (with some variations, an expression of this kind introducing the particular claim in need of proof is nearly universal in Greek mathematics). In short, Greek mathematics is not merely a genre like all other genres, bound by its own rules: it is the most regimented form of writing in all of Greek literature. (In Late Antiquity, the commentator Proclus ruled that each proposition is divided into six parts in the exact same order, a broadly correct observation.) It is very distinctive, indeed, in its very reliance upon *writing*. Greek literature emerged from performances and throughout, Greek readers expected their canonical literary literature to be experienced as spoken words. One could, of course, and did, read silently, but what a papyrus roll was *for*, was the imaginary experience of a play, a dialogue, a speech. I explain this because this accounts for the minimalist nature of the Greek literary roll: it is not embellished or illustrated in any way. Of course, rolls were there so as to be read and the visual vividness was supposed to be provided by performance, not by writing. Mathematics, however, was all inside the roll, no imaginary performance involved and it was about the diagram—present on the papyrus—referred to through letters, labeling the key points on the diagram (which are, of course, a hyper-literate device). Mathematics was marked as the only illustrated genre of writing—which came to reinforce its identity as a genre.

The habits of genre were produced in a context of literary performances. Mathematics, a relatively esoteric field pursued by smaller networks of dedicated experts, took the idea of the genre—and made it literate, and technical. For us, this is obvious (opening a scientific journal, we expect it to be written in a particular, expert, technical manner). This expectation is a historical construct and it was constructed, specifically, by the Greek invention of the idea of a mathematical genre.

When was the invention made? The quote above is from Euclid's *Elements*, a book authored, in all likelihood, very early in the third century BCE. This book is also the earliest extant work in Greek mathematics of any substance. It comes more than a century later than Hippocrates of Chios. For over a century, then, we have no direct evidence of Greek mathematical writings, merely fragments preserved by later authors and testimonies by non-mathematicians. These, however, are very suggestive. Aristotle, for instance (writing in the fourth century BCE), often makes brief allusions to mathematics: they are all broadly consistent with the style we find in Euclid's *Elements*. The *Elements* itself is fundamentally a compilation of past work. It seems that its primary goal was to collect the "toolbox" required for the proof of the more difficult results obtained by past mathematicians. There are hundreds of results in this "toolbox" alone, suggesting that Greek geometers, in the intervening years, have obtained many more difficult results. Indeed, the scattered references we have show that there must have been many dozens of mathematicians, with a marked concentration in the early fourth century BCE. These were the mathematicians who inspired Plato on his mathematically inflected philosophy. They include such names as Archytas, Theaetetus and Eudoxus (among other things, it appears that Archytas found a stereometrical construction solving what, in modern terms, is the finding of a cubic root; Theaetetus classified irrationals; Eudoxus measured the volume of a cone). The oldest and most notable in this group was Archytas (suffice to note that he was something like the political leader of his city Tarentum, perhaps the most significant city of his time in southern Italy). If anyone had the social authority to become the example, followed by others, it would have to be Archytas; conversely, it is hard to see how anyone later than him could have written in a way radically different from that of such a master. Archytas was roughly Plato's coeval, active through the first few decades of the fourth century. In all likelihood, the style we find in Euclid's *Elements*—and in all later Greek mathematical works—was already there, from the time of Archytas if not before. Greek mathematics was an established genre—perhaps, almost from the very beginning.

THE INVENTION OF PROPORTION

Among other things, we noted Archytas discovered a stereometrical construction that found, effectively, a cubic root: we will get to this below. But he was probably more famous for his contributions not to pure geometry but to a different kind of science: mathematical music. It is somewhat surprising to us that this was such a significant part of ancient mathematics (or even that there is such a thing as mathematical music at all). But early on in the fourth century, music became paradigmatic because it brought together mathematics and philosophy—indeed, it brought together, specifically, Archytas and Plato.

The observation is this: certain sounds, heard simultaneously or in quick succession, are pleasing to the ear—they form harmonies. Other combinations sound harsh and unpleasant. They are disharmonious. Late in the fifth century—even before Archytas—the suggestion was made that what constituted a harmonious combination was a simple integer relation. Specifically, the fifth was 3:2 and the fourth was 4:3 (the octave, then, was 2:1). This is indeed evident in a bridged instrument such as the guitar, but it is somewhat mysterious how the Greeks came to notice this, as their stringed instruments were harp-like and were tuned by tension rather than by length. At any rate, Archytas was the first to offer a systematic account of possible structures of the octave. (This is more difficult in the case of Greek music, where more intervals were recognized than in modern music; notice also that, of course, there is no expectation that the gap between consecutive notes would be identical, a condition approximated only in the modern, well-tempered instrument.) Even more significant, Archytas provided a mathematical theory—with proofs—showing why some structures were possible and not others. Specifically, our evidence suggests that he proved that there are no integers n, Q such that $n+1:Q::Q:n$. (hence, for instance, if the octave is indeed 2:1, we immediately find that it cannot be divided into two equal intervals in integers). This is already sophisticated arithmetic, and we get a sense of Archytas' stature, the founder of entire new disciplines of mathematics.

But what mattered to readers such as Plato was not the abstract sophistication of the arithmetic but the metaphysical implication of music theory as such. A concrete phenomenon—the perception of harmony—could be explained through an underlying abstract structure—numbers in proportion. This, to Plato, was paradigmatic of how the world should be understood and while not all mathematicians were Platonists, the discipline of mathematics was approached by many other philosophers and lay readers in antiquity in such terms. Mathematicians, it was understood, found the structures in the world, very often expressed through symmetries and proportions.

Proportion, indeed, is a key theme of Greek mathematics. Already Hippocrates of Chios noted that to double a cube, it will not do the double its side. (You will then get not double the original volume of the cube, but instead eight times that original volume.) He noted that to find the correct length of the side of the cube, you need to consider four lines. Take one line A, whose length is that of the original side of the cube. Then take another line D, whose length is twice the line A. Then, find two more lines B, C such that:

A:B::B:C::C:D

Once you have such four lines in continuous proportion, the line B is the line solving the problem: a cube built on B has twice the volume of the cube built on A (in our terms, if A is 1 and D is 2, B is the cubic root of 2). Hippocrates of Chios himself did not solve the problem of finding four lines in continuous proportion and the first to do so was, indeed, Archytas, offering a very complicated stereometrical solution involving the intersection of a cylinder, a cone, and a torus. Many later Greek mathematicians would offer alternative, more direct solutions—as we will return to note below—and in general this type of problem would become typical of Greek mathematics. Problems were often stated in the vocabulary of proportion (find terms, standing in given ratios to each other); and then solved through the techniques of geometry (configuration produced, so that lines stand to each other as the terms required). Music might have been a significant inspiration but, in the end, there are only so many proportions one can find through integers. The great bulk of Greek work in mathematics involved, specifically, geometrical proportion. In the following section, we note a key example.

THE INVENTION OF ANALYSIS

Greek mathematics became early on a rigidly defined genre. This evolved together with a set of techniques for writing, and also for finding, mathematical results. One of the most powerful is that of analysis. In this technique, when assigned a task, we begin by assuming it solved and then deduce the preconditions from which this solution could arise. This is best seen through an example—and one of the earliest reported is also one of the most significant. We start, indeed, from the problem of finding four lines in continuous proportion: with A, E given, find B, G such that A:B::B:G::G:E. (To recall: in our terms, with A scaled to 1, B is the cubic root of E.) Archytas offered a very difficult three-dimensional solution. A couple of generations later, Menaechmus found a much more direct approach, based on an analysis. (The original text—written just a little prior to Euclid—is lost, but it is cited in full by later mathematicians and in what follows I assume that this is a fair citation, close to Menaechmus' original.)

Let us assume, then, A:B::B:G::G:E. First of all, focus on A:B::G:E. Proportions can be transformed into shapes and so A:B::G:E implies that the two rectangles—the one contained by A, E, and the one contained B, G—are equal to each other. This is interesting, since the lines A, E—the extremes of this proportion—are the "given," or known elements. This means that we also know the magnitude of the rectangle contained by the two unknown lines B, G. By the same logic, we can also conclude that the rectangle contained by A, G is equal to the square on B. Once again, A is a given line: and so what we learn is that B, G, have an interesting internal structure. The rectangle contained by G and a given line (namely, A), is equal to the square on B. B scales, we find, as the square on G.

By analysis, then, we found that the condition A:B::B:G::G:E (A, E given) is satisfied when the two lines B, G enclose a rectangle of a certain fixed area; and when they scale in a particular way: B scaling as the square on G.

To begin with, as in the diagram 3.4, the rectangle KQZD is of a fixed area, specifically it is assumed to be equal to the rectangle contained by the two lines (shown on the side) A, E.

This rectangle may be imagined as a transformable rectangle, by moving the point Q along a curve, so that the two sides of the rectangle always remain parallel to KD, DZ. (This curve obviously never touches the two lines, KD, DZ, since the rectangle's area is fixed and since, were the curve to touch KD or DZ, the rectangle's area would then become zero.) For the time being, we simply postulate the existence of such a curve, which we may call the-equal-rectangle-preserver.

Further, the square on the line QZ is assumed equal to a rectangle contained by the line DZ, and the fixed line A It is assumed then that there exists another

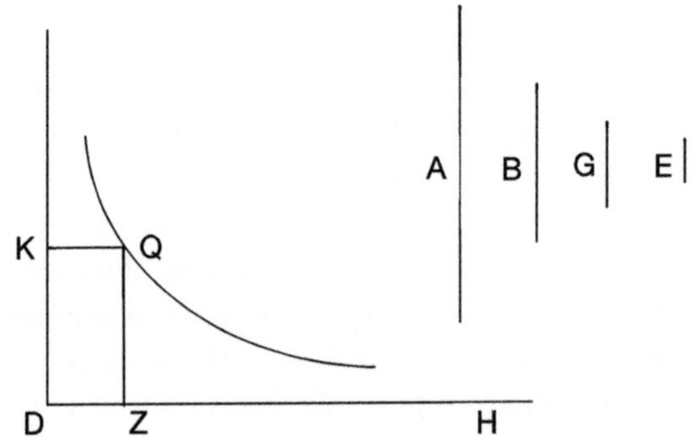

DIAGRAM 3.4. The "equal-rectangle-preserver."

curve, such as that drawn through Q and D, with the property that for any point Q on the curve the square on QZ is equal to the rectangle contained by ZD and A. As before, we simply postulate for the time being the existence of such a curve, which we may call the-square-to-line-scaler. In modern symbolism as in Diagram 3.5.

$QZ^2 = ZD*A$

$Q'Z'^2 = Z'D*A$

QZ may be varied, along this curve, together with DZ. Whichever variation of QZ we get, squared, will always scale with which variation of DZ we get, the line. For the time being, we simply postulate the existence of such a curve, which we may call the-square-to-line-scaler.

The idea now is clear. In the terms of the diagram above, we are given A and E and are asked to find B, G so that A:B::B:G::G:E. This is how we proceed, then. We draw two orthogonal lines. We draw the-equal-rectangles-preserver between them, so that the rectangle it preserved is equal to the one contained by the two given lines A, E. We draw the-square-to-line-scaler through the intersection of the two orthogonal lines, so that the squares it obtains will be equal to a rectangle whose fixed side is equal to A.

The point at which the-equal-rectangle-preserver intersects the-square-to-line-scaler solves the problem. We draw everything, find the intersection Q and complete the rectangle; obviously, the two mean proportionals between A, E must be the lines KD, DZ.

This method of analysis is heuristically powerful and mathematicians often have recourse to it today (it is, for instance, among the heuristic techniques recommended by Pólya in his *How to Solve It*). Quite simply, the more the mind is furnished with, the more it has to work with: and if you assume the problem already solved, you furnish the mind with something extra to work

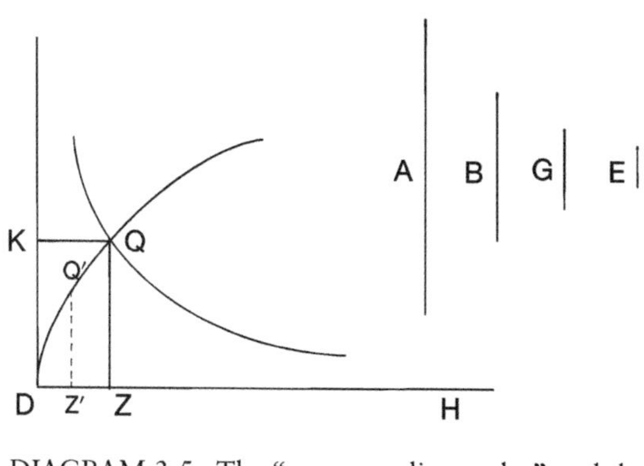

DIAGRAM 3.5. The "square-to-line-scaler" and the "equal-rectangle-preserver."

with. Note, however, that the method is not anything like an algorithm for finding proofs (one still needs to find, *creatively*, the obtainable pre-conditions). Note also that this method is quite different from the "analysis" you learned in high school. (How Greek analysis became modern analysis—by first becoming early modern algebra—is a major question, covered by later volumes of this history.) Note, finally, that even if, no doubt, analysis played a role in how the Greeks discovered their proofs, none of the many dozens of analyses now extant from antiquity comes from the context of discovery. Since they are extant, they come, by definition, from the context of presentation. Indeed, it is rare for analysis to be published alone, and the typical thing is to present the analysis, followed by a synthesis. For instance, in a synthesis of the problem above, you begin by drawing the two curves and then show that the resulting lines solve the problem of four mean proportionals. A synthesis, in principle, renders its preceding analysis superfluous, which raises a question: why did Greek mathematicians publish their analyses at all? (To clarify: in the less frequent, but common enough cases that they do. Most times, analyses were *not* published.) The answer to this must be speculative but what must be important is that an analysis makes the construction it discovers appear natural. Following the analysis above, it becomes clear that the combination of the-equal-rectangle-preserver and the-square-to-line-scaler is a very plausible way of solving the problem of finding four lines in continuous proportion. This makes Menaechmus' solution much more attractive and compelling even though, otherwise, his two curves might appear contrived.

Indeed, in this case analysis led to a major discovery. Whether by Menaechmus himself, at the same time that he offered his original solution, or by some other Greek mathematician, working independently, it was realized, late in the fourth century, that the the-equal-rectangle-preserver and the-square-to-line-scaler occur, so to speak, "in nature." They are not mere mathematical constructs, useful for the solution of a particular problem. They can be constructed in a fairly natural way. Specifically, they are both *conic sections* (and would eventually become canonized through Apollonius' of Perga major work, the Conics, on which more below). Cut the surface of a cone in a certain way and you get what we now call a "hyperbola." This is, in fact, the-equal-rectangle-preserver. (All rectangles intercepted between a hyperbola and its asymptotes are equal to each other.) Cut the surface in a different way, and you get what we now call a "parabola"—and this is, in fact, the-square-to-line-scaler. (Any line drawn "ordinate-wise" from the curve the parabola, on its diameter, scales, in square, as the line segment it intercepts from the diameter.) Yet a third way of cutting the conic surface yields what we now call the "ellipse," which also produces its own distinctive proportions. In all cases, we can say, curves found on surfaces of a cone produce an interesting array of relations and proportions. Be that as it may, at some point late in the fourth century BCE, the Greeks—probably, working through the method of analysis—came across the properties of the conic sections. Conic sections will ultimately be found both in Galileo's projectile trajectories and in Kepler's planetary orbits, which Newton, working backwards, would reduce—through his own analysis—to the inverse-square law of universal gravity. In a very real way, modern science could not come about without the Greek discovery of conic sections: specifically, the analytic discovery that certain proportions imply specific geometrical figures.

THE INVENTION OF AN AUTONOMOUS SCIENCE

About a century later than Menaechmus—probably in the 240s—a young man sent a letter to his friend, the astronomer Conon. In this letter he announced a remarkable array of discoveries: the volume and surface of a sphere, the area contained by a spiral, the volumes of parabolic segments, rotated around their axis . . . The letter merely announced the results without providing their proofs, because it was a challenge, meant not for Conon but for anyone around the Greek Mediterranean. The author, an upstart, made a claim for fame: can you show yourself as the equal to me, Archimedes?

There are many things we know about this letter, even though it is now lost. In his future writings, Archimedes would often refer to it, as he would gradually reveal the proofs to the original claims. Those later writings also reveal, to us, certain facts.

First, no one had answered Archimedes. The most likely explanation is that there were but few mathematicians at the time. There was considerable activity in the generation of Archytas, early in the fourth century BCE, but already Menaechmus might have been among the last authors in this flurry of writing. Euclid summed up the work of this generation but contributed relatively less of his own and in those years, late in the fourth century and through the first half of the third century, mathematical writing became less common, and authors inspired by the mathematical tradition usually contributed not to pure geometry but to astronomy. Archimedes wrote to Conon, an astronomer, after all.

Second, Archimedes sent his original challenge specifically hoping to trip up would-be challengers. In one of the later writings, Archimedes reveals, retroactively, that two of the claims made in the original letter were false, and slyly marked as such. The plan was for people to claim that they proved the results—and then to be shown up as empty braggarts. No one fell for this, because no one tried. But the point is to understand the sense of a tournament, playful but with a rather vicious edge.

Everything written by Archimedes—and he has written dozens of works, of which at least ten still survive in full—is about maximally surprising results. He measures the unmeasurable, he proves the seemingly unprovable. There is an air of whimsy to quite a few of his works. A work now surviving only in fragmentary form contains a counting of the ways of setting up a certain child's puzzle (the Stomachion). Another work, merely attested, counted the faces, sides, and angles of all the semi-regular solids, strange, and unwieldy figures. In the Sand-Reckoner, Archimedes counted the number of grains of sand it would take to fill the entire universe. A lot, indeed, is about surprisingly big numbers, producing what I called in Netz (2009) a carnival of calculation.

Many of the objects studied by Archimedes are strange, riveting, and riveting in their strangeness. The beauty of the spiral is undeniable—and it is incredible, in fact, that a certain straight line, produced by a tangent to a spiral, is found equal to a circumference of a given circle; that a certain figure, defined by the same spiral, is, in area, exactly one third of the same circle. In *Spiral Lines*, Archimedes comes closest to the squaring of the circle—the original inspiration driving already Hippocrates of Chios. This is in the background of much of Archimedes' best work: the volume and surface of the sphere, as noted, but, above all, results having to do with objects arising from the conic sections. Indeed, these seem to be the focus even of his apparently physical works. Archimedes' *Planes in Equilibrium* proved the law of the balance—and then used it to find the center of gravity of a parabolic segment. Archimedes' *Floating Bodies* proved the law of buoyancy—and then used it to find the condition of stability of a segment of a paraboloid of revolution, immersed in a liquid.

Archimedes may have started out in a fairly sparse mathematical landscape. Perhaps his very presence sparked a lively response, and it appears that late in the third century, the Mediterranean was abuzz with new mathematical contributions, largely "Archimedean" in character. Chief among those new authors was Apollonius of Perga, as brilliant and prolific as Archimedes himself but with only one work still extant in Greek. This one work, however, is the *Conics*, the central treatise on its topic, with four books setting out a deep set of results in elementary conic theory followed by four more books containing specific developments of the subject. Of the latter, the fifth, in particular, is spectacularly sophisticated. (It studies the conditions under which a line drawn from a conic curve, onto the diameter, is the shortest of the lines drawn from that point on the diameter, to the curve.) Many emulate Archimedes' subtlety, surprise, and manner of bringing together different realms. So, for instance, just as Archimedes developed results such as the law of the balance or the law of buoyancy, so as to operationalize them into sophisticated geometrical results, so authors, in his wake, found ways to operationalize optics. It is typical that the key result, in this tradition (now known through Diocles' *Burning Mirrors*, a treatise extant only in Arabic translation), concerns the same parabola with which we are so familiar. Diocles has shown that if a mirror is in the shape of a parabola turned around its axis, and it is directed at the sun (so that the rays of the sun become parallel to the axis), then all the rays will be reflected to a single internal point, on the axis. At this point, then, the parabola can cause a fire to burn.

None of this is practical. Later tales have Archimedes turn mirrors on the Roman invaders (who would sack his native Syracuse, and kill him, now an old man, in 212 BCE). But these are legends. Archimedes probably did work on the catapults that defended his city, but this had little to do with mathematical theory. The most evident feature of the mathematical theory of his generation was the emphasis on flights of fancy, surprising and whimsical objects, above all the contrived, ad hoc constructions of the conic sections. A century before, Greeks have discovered that conic sections are useful for solving problems such as the doubling of the cube. In many other traditions, this would be the end of it: the conic sections would have been kept as the stepping stones for later applications, perhaps, but would not be studied as an end in themselves. The typical thing about the generation of Archimedes—a feature often revived in later iterations of mathematics—is that tools become objects. Why? Because mathematical technique, as such, is seen as interesting in and of itself. Science, autonomous.

Why this had happened in the generation of Archimedes has little to do with the kind of autonomy scientists now enjoy in their academic departments. Archimedes had no tenure. However, he wrote for an audience that seems to have valued surprise and originality above practical applications. It is tempting to connect this to wider features of the culture of the time. This was the civilization of the new courts such as that of the Ptolemies of Alexandria. Erudition and sophistication were valued in such courts, cultivating the memory of Greek culture. The scholars marked rare words in Homer; and Archimedes invented the spiral lines.

There is much more than one can add, of course, for the history of Greek mathematics, and which we put aside in this brief survey. Later authors, less easily classified into generations, did make their own contributions, of which perhaps the most significant is that of Diophantus whose solutions to arithmetical problems were influential for the rise of algebra in early modern Europe. But the emphasis on just these two generations—that of Archytas, and then that of Archimedes—remains valid. Over two periods of about fifty years each, the Greek-speaking world has created the path for future mathematics: inventing the author, the genre, the centrality of proportion and the tool of analysis, and the very idea of an autonomous science.

CHAPTER FOUR

Mathematics and Worldviews

The Distinctiveness and Diversity of Greco-Roman Mathematics and the Visions of the World It Underpinned

G. E. R. LLOYD

Our term "mathematics" derives from the Greek *mathēma* to be sure, but that was a general word for any branch of learning. We can, of course, recognize certain investigations in the areas of what we call arithmetic, geometry, harmonics, mechanics, and astronomy that both we moderns and the ancients themselves considered part of "mathematics." Yet, the assumptions of the ancient investigators were sometimes very different from those we normally make, and indeed they differed among themselves.

The interactions between "mathematicians" and "cosmologists" are further complicated by the fact that neither dealt with an agreed and well-defined subject-area. In any case many individuals had interests that spanned both general fields. In Greece, and not only in Greece among ancient civilizations, many people were sufficiently impressed by the prestige and power of "mathematics" to see it as in some sense the key to the understanding of the world in general. Yet, the way in which that was imagined differed profoundly. In the Greco-Roman world, what came to impress many commentators was the way in which mathematics can deliver certainty. If the primary axioms are self-evident and the inferences from them valid, then the conclusions are incontrovertible. Yet, the notion and practice of axiomatic-deductive demonstration were not a necessary condition for the preeminent prestige of mathematics, as we can see from other ancient civilizations that lacked any such concept and yet engaged in sophisticated mathematical investigations. Ancient Egypt, Mesopotamia, and India all exemplify that, and another ancient culture, China, where again there is no equivalent to axiomatic-deductive demonstration, provides texts that explicitly proclaim that mathematics enables one to solve many puzzling questions. It is possible, for instance, to determine the height of distant objects by taking angular measurements from two positions at a given distance and using similar right angled triangles to get the result. In a text that dates from the turn of the millennium, the *Zhoubi suanjing*, this technique was even used in the context of assessing the height of the sun—on the assumption of a flat earth. This will be the subject of detailed discussion in a subsequent volume, but I include a translation of the key text in the Annex here.

The ways in which mathematics is brought to bear on worldviews differ not just as between different ancient civilizations, but also within the Greco-Roman one. The topics on which different Greek and Roman writers disagreed and on which they engaged in vigorous debate include the following fundamental issues. (1) What does mathematics study? Are there separate mathematical objects and if so how do these relate to the rest of the phenomena of experience? (2) Do different branches of mathematics have indeed different such objects? How do the objects of "arithmetic" relate to those of "geometry," not to mention those of such "applied" studies as "harmonics," "astronomy," statics, dynamics, mechanics and cartography? (3) What modes of cognition does mathematics secure? How far can a model of axiomatic-deductive demonstration be reasonably applied? (4) What understanding of natural phenomena and the cosmos as a whole, including of the place of humans in it, can mathematics yield? While a mathematical approach gives robust-seeming results over a certain range, other areas of experience appear to resist such an approach. (5) How far do the claims for mathematics as the best or even the only sound method have to be qualified in the light of the difficulties it encounters? For many, mathematics was good for more than just knowledge and understanding. By providing access to the profoundest truths, it contributed to happiness and fulfilment. On that view, this discipline had moral and even religious implications. Training in mathematics was compared with, even seen as an example of, a kind of initiation, whose aim was to reveal the mysteries of the universe. Yet, once again this point of view was contested. We shall need to bear these five issues in mind as we tackle some of the rich, if often problematic, evidence in our sources.

THE EVIDENCE FOR PYTHAGORAS, PHILOLAUS, AND ARCHYTAS

We may begin with the Pythagoreans where immediately we face major difficulties in the evaluation of our evidence (Huffman 2014). We know that Pythagoras of Samos was a historical figure dating from the late sixth century BCE since he is referred to in fifth-century sources, Heraclitus and Herodotus for instance. We hear from Plato (*Republic* 600b) that he taught a way of life. The trouble is that in the revival of an interest in his philosophy from the second century BCE to the third century CE what passed as his teaching and achievements became more and more the stuff of legend. What exactly may be reliably attributed to him by way of mathematical discoveries and activities? Many different proofs of the theorem named after him are to be found in our Greek sources, and we are in no position to identify the author of the one we are given in Euclid's *Elements* (I.47) (Figure 4.1). Yet, we know that already many centuries previously the Babylonians were well aware of so-called Pythagorean triplets, numbers such that the squares of the first two equal the square of the third (Plimpton 322; see Chapter 3). True, this knowledge was not accompanied in our extant Mesopotamian sources by a geometrical proof. Yet, it is clear that they were aware not just of that particular set of triads, but of the general properties they all exemplify.

Then Pythagoras is associated by Greek writers with the discovery that the major concords of octave fifth and fourth can be expressed as ratios between integers, 2:1, 3:2, 4:3. Yet, it is extremely implausible to think that practicing musicians were not well aware that a fifth and a fourth together constitute an octave whatever representation they may have adopted of those relations.

One of the major modern authorities on Pythagoreanism (Burkert 1972) has accordingly argued that the mathematical activities associated with Pythagoras himself should more correctly to assigned to a group of his followers who came to be labeled "mathematicians" indeed, starting in the fourth century BCE. Pythagoras' own teaching, on Burkert's view, related rather to his doctrine of the transmigration of souls and to certain moral or ritual precepts (akousmata, the legacy taken up by the second group of his followers who went by the name of "akousmatikoi").

Yet, while a healthy skepticism is in order with regard to attempts to reconstruct many aspects of the life and work of Pythagoras himself, we do have some reliable (as well as a fair amount of shaky) evidence concerning two important fifth-century followers, Philolaus and Archytas, and this provides a far sounder basis on which to begin our exploration of the themes of this chapter. To that we can add the detailed account of views that Aristotle ascribes

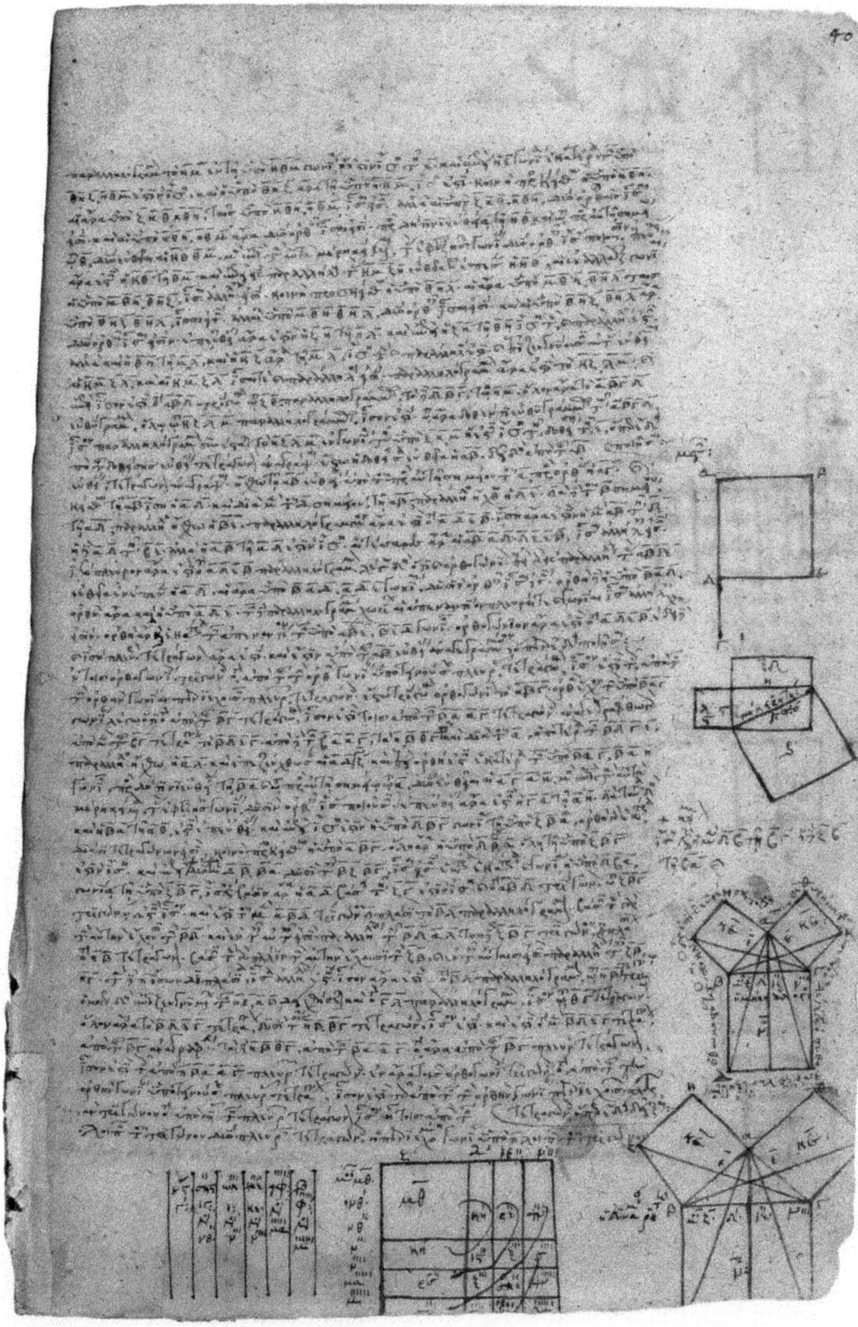

FIGURE 4.1. The Pythagorean theorem in a medieval manuscript. Biblioteca Apostolica Vaticana, Vaticanus Palatinus graecus 95, fol. 40r. Photo by Wikimedia Commons.

not to Pythagoras himself but to some who were associated with him as his followers, although again we are dealing with fifth rather than sixth century work. For our purposes here, the questions of the precise relationship of these individuals or groups to Pythagoras or others who came to be labeled Pythagoreans is less significant, for it is not the philosophical affiliation that is important for our purposes so much as the content of the views reported.

Both Philolaus and Archytas make pronouncements on the relationship between different mathematical disciplines (my question 2 above). Both see "mathematics" as covering a variety of studies, number-theory, geometry, harmonics and the study of the heavens, yet they take different views of the hierarchy within those. For Philolaus (Fr. 7a, cited by Plutarch, *Quaest. conv. 718e*, Huffman 1993: 193ff.) geometry is the "source and mother-city" of the rest of the mathematical sciences, while Archytas (Fr. 4 from Stobaeus, Huffman 2005: 225ff.) held that "logistic" (calculation, number theory) is superior to all the others, geometry included. This might seem to be a mere matter of taste or to reflect some sense of which area of mathematics was better developed. But if we consider the usual Greek assumptions about number, a more important ontological issue is at stake. For most Greeks, numbers (*arithmoi*) are integers greater than one. So, one itself is not a number, and fractions are understood as ratios between integers. More importantly, the numbers form a discrete series. By contrast, geometry (on most views) deals with a continuum: the geometrical line unlike the numbers series is infinitely divisible. The fundamental problem of how to reconcile these two views runs through Greek mathematics and surfaces again in harmonic theory, where some writers such as Aristoxenus (*fl.* 335 BCE) treat sound as a continuum while others insist that concords must be expressible as ratios between integers.

Philolaus' expressed preference for geometry may also be thought to clash with the sophisticated general metaphysics he developed based on a contrast between "limiters" and "unlimiteds." The distinction is an obscure one whose interpretation continues to be disputed: on one view "limiters" would be exemplified by shapes and structures and "unlimiteds" to what is so shaped and structured. This in turn was associated in some way with a classification of numbers based on odd and even integers (a third kind is a mixture of the two, just as he also acknowledged mixtures of "limiters" and "unlimiteds"). While much remains unclear, it appears that he was working towards an epistemology or a methodology according to which number provides the basis of a claim to have knowledge of any group of phenomena. Thus, one fragment (4 from Stobaeus, Huffman 1993: 172ff.) puts it that "all the things that are known have number. For it is not possible that anything whatsoever be understood or known without this."

WAS THERE A FOUNDATION CRISIS IN GREEK MATHEMATICS?

Whatever reconciliation was attempted within Philolaus' views and between them and Archytas', the idea that there was some foundation crisis in Pythagorean mathematics is not borne out by the facts. The ancient legend taken up by some positivist historians of mathematics (Hasse and Scholz 1928) had it that the discovery of the incommensurability of the side and the diagonal of the square led to a collapse of the Pythagorean program to use number as the key to the understanding of phenomena. The discoverer, sometimes identified as Hippasus, is said in our sources either to have drowned at sea or to have otherwise been punished for the scandalous revelation (Iamblichus *Life of Pythagoras* 88 and 247, DK 18A4). But for the Greek mathematicians themselves, a geometrical treatment of the problem of that incommensurability may have proved an adequate solution. After all, the square on the diagonal is equal to the squares on the other two sides: so, incommensurability in number is perfectly compatible with commensurability in square, that is, even when lengths themselves are incommensurable, it may be that the *squares* on a pair of lengths *will* have the ratio of a number to a number.

In practice, there are no signs of any waning in mathematical activity in the late fifth century, when many others besides Pythagoreans are known to have been involved. A variety of writers with many different philosophical allegiances or with none was involved in cracking one or more of the three special problems, of squaring the circle, trisecting an angle and duplicating the cube (Knorr 1986). This last was the subject of a particularly sophisticated study by Archytas, which involved a complex geometrical construction determining a point by the intersection of three surfaces of revolution, a right cone, a cylinder and a tore (Eutocius citing Eudemus, and Huffman 2005: 344ff.).

But in other cases other mathematicians not associated with the Pythagoreans were involved. "Squaring" a curvilinear area meant finding a non-curvilinear area equal to it. Our very first reliable piece of extended mathematical reasoning tackles one particular example of this. This is the attack on the quadrature of lunes that is ascribed to Hippocrates of Chios (Simplicius *In Ph.* 55.25ff.), as discussed in Chapter 3. Hippocrates may also have been responsible for the first attempt at a systematization of mathematical knowledge (Proclus *In Euc.*2.66.4). How far indeed his writings anticipated Euclid's *Elements* is not clear: we are not even certain that *Elements* was the title of his work, as opposed to one given to it by later commentators who saw it as a precursor to Euclid. Yet, his work on quadratures certainly shows great confidence in both the geometrical methods used and the results obtained.

On the basis of what we know of fifth-century mathematical studies it would be more plausible to argue that the discovery of the incommensurability of the

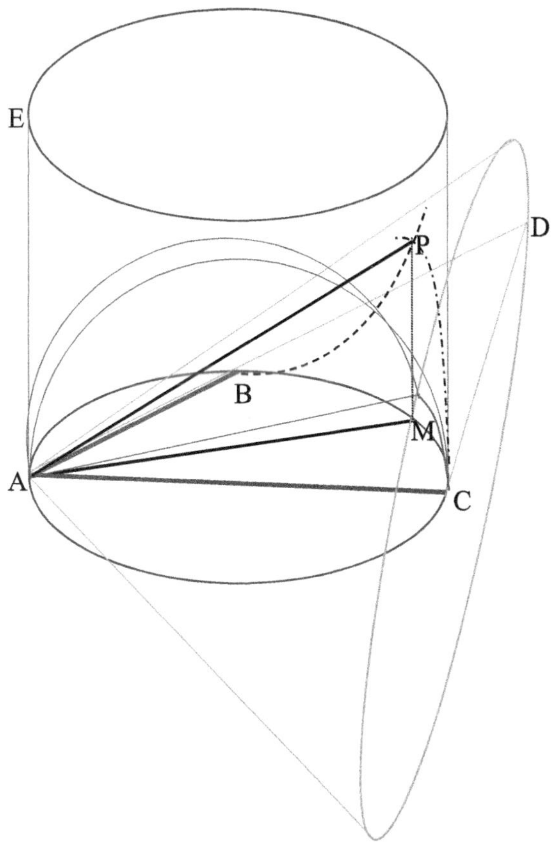

DIAGRAM 4.1. Archytus' complex construction for doubling the cube. There are three intersecting surfaces: a cylinder with base circle ABC; a cone with axis AC and base circumference CD; a half torus produced by rotating the half circle on AC perpendicular to the base of the cylinder about the line AE. The cone traces out the curve BP on the cylinder, while the torus traces out the curve CP; where they meet at P provides the solution of the problem, namely, it turns out that PA and AM satisfy the proportion AC:AP::AP:AM::AM:AB. In particular, if AC=2 and AB=1 then $AM^3/AB^3 = AM^3=2$.

diagonal and the side of the square acted as a stimulus to research rather than as a factor that inhibited it. That result came to be the subject of a rigorous proof, one example of which we find in Euclid himself which proceeds by reductio ad absurdum (*Elements* Book X, app. 27). If the diagonal and side are indeed commensurable, then the same number is both odd and even. That would be a self-contradiction: so the diagonal and side must be incommensurable. That was only one of many demonstrations of this result. However, the important point for us is that all such strict proofs exhibited the power of mathematics rather than its weakness.

THE ARISTOTELIAN ACCOUNT

We have yet to consider the important testimony of Aristotle on Pythagoreans whom he neither names nor securely dates. Just after he has been dealing with Empedocles, Anaxagoras and the atomists, all thinkers active in the late fifth century BCE, he has this to say:

> Contemporaneously with these philosophers, and before them, the so-called Pythagoreans, who were the first to engage in mathematics, advanced this study, and being trained in it, they thought its principles were the principles of all things. But of these principles numbers are by nature the first, and in numbers they seemed to see many resemblances to the things that are and come to be—more than in fire and earth and water . . . And again they saw that the modifications and the ratios of the musical scale were expressible in numbers. Therefore, since all other things seemed in their whole nature to be modelled on numbers, and numbers seemed to be the first things in the whole of nature, they supposed the elements of numbers to be the elements of all things, and the whole heaven to be a musical scale and number. (*Metaphysics* 985b23ff., trans. by the author)

Since this is one of the richest and most telling texts for the purposes of our investigation, we should pause to comment on several of its features. First, the context in which it comes is Aristotle's examination of the principles, *archai*, that his predecessors had invoked, where he wants to show that his own view of the so-called four causes, material, formal, efficient, and final, comprehends and supersedes all earlier opinions on the subject. Aristotle has just been interpreting the earliest Milesian philosophers as being concerned with the material cause. Thales' "water" is, in Aristotle's view, the material cause of everything else, the stuff of which they are made, although, of course, Thales himself had no term for "matter" and it may well be that he was more interested in how the world originates than in what it is now made of. Indeed, the word Aristotle uses for "principle," *archē*, also means "beginning."

So, when Aristotle reports that the Pythagoreans chose numbers rather than fire or earth or water as their principles and illustrates this with the example of musical scales being expressible in numbers, there is a tacit contrast between a material and a more formal account of reality. This would suggest that appeal to mathematics produced an important change in the way some viewed the world. Yet, to go too far along that line of interpretation is clearly a mistake and not just because the vocabulary of matter and form was not available to the Pythagoreans any more than it was to the Milesians. Other reports have it that for the Pythagoreans, or at least for some of them, numbers are not just what Aristotle would call the formal principles of things, but indeed what they are made of, numbers themselves being conceived as concrete material objects. According to a text in Theophrastus (*Metaphysics* 6a) which cites Archytas, no less, a Pythagorean named Eurytus is said to have assigned particular numbers to particular kinds of things, in each case the number being the minimum amount of points by which the outline of the kind in question could be represented.

If we ask the first of my five questions about the early Pythagoreans ("what kind of objects does mathematics study?") it is as well to recognize that the evidence hardly permits a determinate answer. Sometimes other things are said to *be* numbers, sometimes they are said to resemble or *be like* numbers (contrast, e.g., Aristotle *Metaphysics* 987b27–8 with 987b21–2). The first of those formulations tends to suggest an ontology in which mathematical objects are the ultimate reality, to which other perceptible phenomena are somehow to be reduced—a view we shall see is developed in Plato. The second is compatible with a far weaker position, where perceptible phenomena exhibit mathematical relations without being reduced to them.

When we consider the various areas of mathematical inquiry that engaged the Pythagoreans as they "advanced its study" in Aristotle's words, we can see that it is important to keep different options concerning the underlying ontology open. Two prime examples are musical harmonies and astronomy. As we have already noted, the fact that the major concords are expressible in terms of ratios served as an important example, and not just for Pythagoreans, nor even just for ancient Greeks, of regularities that could be revealed by a mathematical analysis. Sound might seem to have no obvious connection with numbers, but concordant ones certainly did and could, in a sense, be explained and understood in terms of the mathematical relations they exhibited, though, as already remarked, this was not the end of the matter. Whether musical sound should be represented as a continuum (like the geometrical line) or consisted in discrete units (like numbers for Greeks) remained controversial throughout Greek music theory (Barker 1989).

But then a second area of prime interest is astronomy, where the eventual ambition of many Greek theorists was to provide geometrical models by which the movements of the heavenly bodies could be explained. Once again, this is a distinctive feature of some Greek speculations, for although the Babylonians especially engaged in sustained investigations into the movements of the sun, moon, and planets (Rochberg 2004), they made no attempt to give those movements geometrical representations in order to prove their orderliness. The chief problem, in that Greek program, was to account for the apparent irregularities in the motions of the sun, moon, and especially the planets, where, in the last case, their stations and retrogradations were what gave them their Greek name, literally the "wanderers." A variety of such models was tried out, concentric spheres, later epicycles and eccentrics, all of which aimed to show that perfectly regular circular motions could be combined to yield the apparent irregularities of the observed movements (see Chapter 5 for illustrations). Those movements then became not so much a scandal for science and cosmology as marvelous evidence of their success, that is in their ability to demonstrate the order of the universe, a primary preoccupation of many Greeks but not of their near Eastern predecessors.

Yet, for the Pythagoreans whom Aristotle discusses, the way that "numbers" were brought to bear on astronomical phenomena was very different. A highly critical report in the *Metaphysics* (986a3ff.) makes the point:

> All the properties of numbers and scales they could show to agree with the attributes and parts and the whole arrangement of the heavens, they collected and fitted into their scheme: and if there was a gap anywhere, they readily made additions so as to make their whole theory consistent. For example, as the number ten is thought to be perfect and to comprise the whole nature of numbers, they say that the bodies that move through the heavens are ten. But as the visible bodies are only nine [that is the sphere of the fixed stars counted as one, plus the five visible planets, sun, moon and earth] to meet this they invented a tenth, the "counter-earth."

This Pythagorean scheme would thus be a prime example of the world being made to fit mathematics rather than using mathematics just to describe the world.

It was indeed not the only speculative aspect of their system. In his *On the Heavens* Book II, ch. 9, Aristotle associates the Pythagoreans with a doctrine of the harmony of the celestial spheres. According to this idea each sphere emitted a sound—for how could such massive bodies move soundlessly? To get round the obvious difficulty that we do not hear any such sounds, they argued that we have become habituated to them from birth. But they further speculated that each of the heavenly bodies makes a different note, that these notes vary according to the speeds of the planet which in turn vary with their distances, although Aristotle does not report how these were determined. The upshot, however, was that the sounds of the heavenly bodies as they moved are concordant. Aristotle describes such a view ironically as "fitting" and "musical," but he would have none of it. The heavenly bodies make no noise since they are passively carried round on their spheres. What makes a noise is what moves in a stationary medium: but the support of the heavenly bodies is not stationary.

Aristotle comes back to attack Pythagorean speculations again in *On the Heavens* (293a18ff.) There he reports that certain Pythagoreans, who according to later sources (e.g., Aetius 2.7.7, Huffman 1993: 237), included Philolaus, held that the central position in the universe was held not by the earth but by what they called the Central Fire. We are not in a position to offer a confident interpretation of the role this played in their astronomical system as a whole, though we have enough evidence to suggest that he was attempting to produce a detailed geometrical model to explain the movements of the sun, moon and planets. But Aristotle's scathing remarks suggest that the motivation for this first non-geocentric system was driven in part at least by metaphysical and religious reasons. Earth is not noble enough to occupy the central position: in Philolaus (Fr. 7, Stobaeus, Huffman 1993: 226ff.), the Central Fire was the first thing to be formed in the cosmogonic process by which Limiters came to interact with the Unlimiteds.

According to Aristotle, therefore, the Pythagorean astronomy contained fantastical ideas that corresponded to certain a priori assumptions and were entirely unjustified by, and indeed in conflict with, the empirical evidence. Yet, this condemnation is certainly at points unfair, as indeed another text in Aristotle goes to show. Although the details of their astronomy are unclear, Aristotle's report in *On the Heavens* 293b23ff. has it that the "counter-earth" was used to explain the higher frequency of lunar as opposed to solar eclipses, as observed from any given location on the earth.

PYTHAGOREAN MATHEMATICS APPLIED

The details of much Pythagorean mathematical work elude us. Yet, an evident ambition to apply mathematics to the understanding of phenomena runs through their harmonics, their astronomy and cosmology. That is, however, far from the end of the story. Aristotle had said that "the whole heaven is a musical scale (*harmonia*) and a number." "Concordance" is not just the feature of reality that allows it to be known and understood. The ethical, moral, even religious associations of harmony are an important element in the mathematical program. The number four, the first square, we are told, represents justice (e.g., Alexander, *In Metaph.* 38.10–16). Opportunity, we also learn, was associated with the number seven and the various phenomena associated with that number once again drew Aristotle's condemnation.

"Why need these numbers be causes?" he asks in *Metaphysics* 1093a13ff.:

> There are seven vowels, the scale exists of seven strings, the Pleiades are seven, at seven animals lose their teeth (at least some do, but others do not) and the champions that fought against Thebes were seven. But is it because the number is the number it is that the champions were seven and the Pleiades consist of seven stars? Surely the champions were seven because there were seven gates or for some other reason and the Pleiades we count as seven and the Bear as twelve, while other folk count more stars in both . . . These people are like the old-fashioned Homeric scholars, who see small resemblances but neglect great ones. (Ibid.)

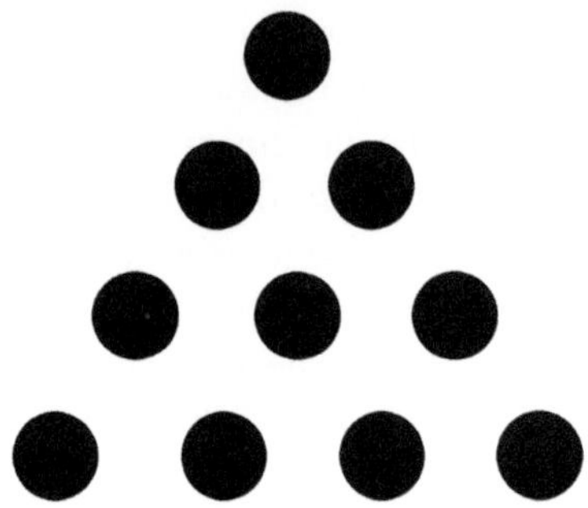

DIAGRAM 4.2. The Pythagorean symbol, the *tetraktys*. Ten represented as 1+2+3+4.

Number symbolism of that type evidently cut no ice with Aristotle, but we find a different way of associating numbers with morality in an explicit statement by Archytas (Fr. 3 from Stobaeus 88.5f and Iamblichus, *On the Common Mathematical Science* 44.10ff., Huffman 2005: 182ff.). "Once calculation was discovered, it stopped discord and increased concord. For people do not want more than their share and equality exists once this has come into being. For by means of calculation we will seek reconciliation in our dealings with others." Obviously, the power of the mere ability to apply calculation to produce peaceable relations is here rather naively exaggerated. Yet, mathematics is no mere intellectual discipline. The number ten is not just complete (as the sum of the first four integers) but holy. In a variety of sources, including for Philolaus (Lucian, *Laps.* 5, Huffman 1993: 355f.) the *tetraktys*, a symbol representing the perfect number ten,[1] was said to be the "greatest oath" by which Pythagoreans swore.

The heavenly bodies are indeed divine, they are living gods, and as we have seen the Central Fire was represented as the Hearth of the Universe. Mathematics thus gives its devotees access to the mysteries of the universe that go far beyond the mere physical explanation of natural phenomena.

We can follow even more proliferations of the applications of the idea that the principles of things are numbers through the various representations of Tables of Opposites that are reported. In Aristotle's version, Limit and the Unlimited figure as the prime opposition (where Philolaus had had Limiters and Unlimiteds) but Odd and Even then lead into a whole series of other contraries on one or other side of the Table, one and plurality, right and left, male and female, at rest and moving, straight and curved, light and darkness, good and evil, square and oblong (Aristotle *Metaphysics* 986a22ff.). Of course, the Pythagoreans were not the only early Greek thinkers to suggest such correlations between pairs of opposites. Aristotle not only goes on to report that Alcmaeon suggested an indeterminate number of such pairs, but even considers that all earlier Greek philosophers adopted opposites as their fundamental principles. The important point to remark, however, is that structuring such oppositions in terms of associations with Limit and Unlimited on the one hand, and Odd and Even on the other gave the Pythagorean doctrine a mathematical underpinning that was absent from many other views. This in turn allowed certain mathematical ideas to be brought to bear in the explanation of a wide variety of physical phenomena. Number, we must recall, is that by which everything is known that can be known.

It is time now to take provisional stock of what we can say about the materials so far reviewed. Advances in various branches of mathematical studies were clearly occurring throughout the fifth and early fourth centuries BCE. We mentioned the attack on specific geometrical problems such as squaring the circle. We saw that Hippocrates of Chios is credited with some early systematization of mathematical knowledge. Our commentators on Euclid's *Elements* tell us that various parts of the book were substantially the work of mathematicians such as Theodorus, Theaetetus, and Eudoxus (Proclus, *In Euc.* 67–8). The last is credited with an influential theory of proportion that applied to incommensurable as well as commensurable magnitudes. We hear of the other two from Plato's dialogue, *Theaetetus*, which reports investigations of instances of incommensurabilities up to that corresponding to the square root of 17.

At the same time what was assumed about the ontological status of the objects of mathematical study (my questions 1 and 2) remained unclear. Numbers may have been represented as the key to the understanding of everything, but precisely how that worked out in practice was anything but agreed. The ethical and moral implications of certain numerical relations, harmonies, and concords are frequently mentioned in our sources. In some cases, in astronomy especially, the objects of mathematical interest were also viewed with genuine religious awe: I shall return to give examples from later astronomers. Mathematics was recognized as being of great practical use, but in certain quarters at least it was coming to be acknowledged as the highest intellectual activity open to humans and a major contributor to well-being and happiness.

THE FOURTH CENTURY BCE: PLATO

The evidence for much of the reconstruction we have thus attempted must be admitted to be far less secure than we should have liked. When we turn to major fourth century figures we are on much firmer ground, not that the ancient controversies that swirl around our five questions are in any way resolved.

Plato's use of mathematics is a case in point. Not himself a practicing mathematician he nevertheless makes heavy use of mathematics both in his metaphysics and epistemology, and in his cosmology. He was, to be sure, not the first Greek philosopher to insist on a contrast between, on the one hand, reason and its findings, and, on the other, sense-perception and its objects. Parmenides before him, especially, had opposed the Way of Truth, based on *logos*, and the Way of Seeming, on which mortals wander deluded by their senses and by their conventional beliefs. But for Plato, this contrast came to form the basis for a far more systematic ontology. True, he nowhere sets out, in so many words, what we call his Theory of Forms sometimes misnamed a Theory of Ideas, though it is clear that the Forms are *not* ideas or concepts, but the true realities. But in many different contexts the speakers in such dialogues as the *Symposium, Phaedo,* and *Republic* contrast Beauty, Goodness, Equality and so on themselves, with examples of beautiful, good or equal things. The latter suffer from a fundamental flaw. What is beautiful in some respect or at some time may be ugly in another or at another. Beauty itself, by contrast, is unqualifiedly beautiful. This does not lead him to suggest that perceptible particulars are unreal. They exist, but the way they can be said to be is qualified. The Forms alone are the true unqualified realities.

Mathematical examples figure prominently in his development of this contrast. A pair of equal sticks may be equal in some respects, but will inevitably be unequal in others. To discover what Equality itself is, we must direct our attention away from perceptible particulars and concentrate on the Form itself. Yet, while mathematical entities are aids in developing our appreciation of the Forms themselves, they do not qualify as among the highest realities themselves, at least according to arguments presented in the *Republic*. There mathematics is praised as a training in abstract thought, but in the famous image of the Divided Line is subject to two shortcomings. On the one hand geometry uses perceptible diagrams, on the other mathematics gives no account of its fundamental hypotheses, here represented by such examples as odd and even and various figures and kinds of angles (*Republic* 510c).

Eventually, though this is not clear in the *Republic*, Plato was said to have developed a notion of what came to be called mathematical intermediates. They share with the Forms the property that they are intelligible, but share with perceptible phenomena that they are plural (where the Forms are said to be unique, *monoeidēs*). So, while mathematics has an important contribution to make to training us to practice abstraction it is still rated below the highest mode of cognition, dialectic itself. But quite how Plato wishes to sustain the claims he makes on its behalf is far from clear. Mathematics, we learnt, fails to give an account of its starting assumptions.

But how does dialectic do so? What kind of account is available to it that is beyond the reach of mathematics? We are told that the summit of intellectual progress is a so-called unhypothesized first beginning, the Form of the Good. But how that is to be reached is left in abeyance. By the time we get to Euclid, indeed already by the time of Aristotle, the status of the axiomatic assumptions on which mathematics in particular was to be based was clarified. The primary propositions must be self-evident and indemonstrable, for if they are capable of being proved, they should be, and then they would not be primary. But obviously that insight into the nature of postulates cannot be applied to Plato's claims, for if we think of his "hypotheses" as axioms, he would appear to be committed to the paradox that dialectic is to be based on an unaxiomatic axiom. Evidently the insights that can lead the philosopher to an appreciation of the Form of the Good must draw on other sources of inspiration than mathematics, a further reason for treating that as lower in the scale of cognitive capacities than his version of pure dialectic.

This is not the only puzzle in the program of the *Republic*. When Socrates discusses the study of harmonics and astronomy in both cases he insists on the mistakes that arise from too great a focus on empirical evidence. At 530d he expresses his agreement with unnamed Pythagoreans that harmonics and astronomy are in some sense "kindred disciplines" (a view found in Archytas Fr 1, Huffman 2005: 103ff.), but he goes on to criticize one way of pursuing the former subject. This consists of attempting to learn about harmonies by studying audible sounds, indeed investigating these experimentally on a string, trying to determine which is the smallest audible interval for example. What they should do is rather to consider which numbers are concordant with others (531c). Similarly, certain astronomers are taken to task for attempting to make progress in their subject by studying the visible heavens. No, he protests (530bc), it is by means of problems that the true potential of astronomy to train us in abstract thought will be realized, and we should "leave the things in the heavens alone."

We have here then a program of a radically anti-empirical practice of the mathematical sciences. That was certainly not the position of those who actually engaged in either harmonics or astronomy, but it serves as a prime example of the deployment of a certain interpretation of mathematics for philosophical or metaphysical ends. Philosophy, we are left in no doubt, is supreme. Mathematics can help us to learn to cultivate our reason. But in Plato's view in the *Republic* the price that has to be paid to achieve that end is to turn away from any attempt to strengthen the empirical basis for the mathematical sciences

Yet, that is not the full story where Plato is concerned. When he came to write a cosmological dialogue, the *Timaeus*, we can detect a certain accommodation to mathematics that goes beyond what we might have expected from the *Republic*. The *Timaeus* offers only a "likely account" of the cosmos, while it still insists (29bc) that where the Forms are concerned, certainty must and can be attained. The chief respect in which Plato's cosmology goes beyond all its predecessors relates to the development of the idea that creation reflects the work of a divine benevolent Craftsman or Demiurge (Sedley 2007). But how does he go about his work of introducing order into preexisting chaotically moving matter? Both the construction of the World-Soul and that of the body of the cosmos draw heavily on mathematical models. The World-Soul is an elaborate arrangement of musical harmonies. And when it comes to what the cosmos is made of, the fundamental elements are represented by combinations of primary triangles, the right-angled isosceles and the half equilateral. The four simple bodies of fire, air, water, and earth, which Plato takes over from Empedocles and others, are not true elements at all. Each is identified with one of the five regular solids, constituted by those primary triangles. Fire is identified with the tetrahedron, air with the octahedron, water with the icosahedron, and earth with the cube. The fifth regular solid, the dodecahedron, is mentioned but not identified with a simple body at *Timaeus* 55c (Figure 4.2).

Thus, as a physical theory we are offered a variation of atomism. But where the original atomists, Leucippus and Democritus, had postulated an indefinite variety of atomic shapes, Plato's version shares the property of being geometrical, but diverges from theirs in insisting on strict regularity. The orderliness of the cosmos was a fundamental tenet and that feature was secured by the basically mathematical character of the intelligence at work in its creation.

Plato's cosmology, even when presented as no more than a "likely account," is breathtakingly speculative and at points appears quite arbitrary. Yet, his moral injunction to cultivate the soul, notably by the practice of abstract reasoning,

FIGURE 4.2. Five regular solids. Photo by SSPL / Getty Images.

and the epistemological basis of his whole philosophy, the contrast between the intelligible and the perceptible worlds, remained immensely influential even while many dissented from other aspects of his teaching. First, those who actually pursued investigations in such fields as harmonics or astronomy generally rejected his recommendations for a purely a priori attack on those subjects. Even harmonic theorists who argued that it was the properties of numbers themselves that determined concords accepted that they needed to study musical sounds. No subsequent practicing Greek astronomer accepted the injunction to "leave the things in the heavens alone," even though differing views were taken about the status of the observable data.

THE FOURTH CENTURY BCE: ARISTOTLE

But then secondly there were those who argued that Plato was quite mistaken on the nature of mathematics and its objects themselves. The first to do so was his pupil Aristotle and this takes me to our next rich source of materials for our study.

Aristotle's philosophy of mathematics has been variously interpreted but it is now generally accepted that it marks a radical departure from Plato (Lear 1982). For Plato, as we have seen, the objects of mathematics are separately existing intelligible objects, differing from the Forms on the one hand, and from perceptible particulars on the other. But for Aristotle what mathematics studies is not such independently existing intelligibles, but rather the mathematical properties of perceptible phenomena. It does so by abstracting from other characteristics of those phenomena, their color or their temperature, for instance. But its focus, nevertheless, is on properties that belong to and are exemplified by perceptible phenomena and that do not exist in isolation from them.

Analogously, where Plato had insisted on the independent existence of the intelligible Forms, for Aristotle what primarily exists independently are individual substances, each of which consists of a combination of form and matter. That immediately reverses Plato's view of where attention should be focused. This should no longer be on independently existing Forms, but rather on the form–matter combinations of the concrete objects of experience. Their intelligibility was not a matter of participating in or imitating a Form, for intelligible forms are now instantiated in composites, not separate from them. Then to suggest that physical substances should be analyzed in basically quantitative, mathematical terms in any of the versions of atomism with which Aristotle is familiar, is to commit a category mistake. The essentially qualitative properties of physical objects, their being hot, cold, wet, and dry, and so on, are not to be reduced to quantitative differentiae, a move that thereby sees off atomism whether in its Democritean or its Platonic mode. The four simple bodies, Fire, Air, Water, and Earth, can be analyzed in terms of combinations of those four primary qualities, fire being hot and dry, air hot and wet, and so on, but any further reductive analysis is ruled out.

That might look as if it torpedoed Plato's whole program of a certain mathematization of physics (to use rather anachronistic terminology) but once again that does not give us the whole story. While not himself engaging in what he called the mathematical branches of physics, harmonics, astronomy, optics, and mechanics, Aristotle nevertheless broadly endorsed the ways in which each of them used mathematics in their study.

The point is clearest in the case of the study of the heavens. In his most notable foray into that field, in book Lambda of the *Metaphysics*, Aristotle reports the theories of the astronomers Eudoxus and Callippus and adds to them on his own account. How far and at what points this work drew on Near Eastern investigations is controversial (cf. Lloyd 1996: ch. 8). Aristotle evidently knew of both Egyptian and Babylonian data and even cites ancient traditions of observation in support of his claim that the heavens are unchanging (*On the Heavens* 270b13ff.). Yet, the agenda that Eudoxus and Callippus set themselves is original. As already remarked, the chief problem was to see how the apparently irregular movements of the sun, moon and planets could be explained, that is seen to be the resultant of the combination of certain perfectly regular circular movements.

For this purpose, first Eudoxus and then Callippus postulated a certain number of concentric spheres, which were successful, at least in qualitative terms, in showing how irregularities could be so reduced, though the actual number of spheres needed remained disputed. But Aristotle's contribution was to deal with the interrelations of these spheres. To prevent the movement of a higher sphere being affected by those below it, he introduced retroactive spheres to cancel out the lower movements. This gave the whole system a concrete physical interpretation, though it should be stressed that Aristotle expresses great hesitancy about the actual number of spheres that will be needed.

His conviction that the whole forms a single coherent system is apparent, even though what that system was made of was not the four elements we see around us on earth but rather a fifth element, *aithēr*, which has the property of moving eternally in a circle, and of being neither hot nor cold, neither dry nor wet. This introduction of a gulf between the so-called sublunary sphere and the heavens themselves has often been condemned as a disaster for astronomy. But to that it must be said that Aristotle thought the conclusion was forced on him by the apparently unchanging circular motion of the heavens which, as noted, he believed to have been established by observations carried out by Egyptians and Babylonians as well as his own Greek predecessors. That movement could not be forced, for then it would not be eternal and unchanging. But if it is natural, it could not be explained in terms of the natural movements of heavy bodies towards the center of the earth (deemed to coincide with the center of the universe) or of light ones away from that center. The irony has often been pointed out that, when judged—anachronistically—from the standpoint of later dynamics, the trouble with Aristotle's account of motion was not excessive abstraction in the case of the fifth element, but rather insufficient abstraction from the effects we would put down to friction in the case of the sublunary ones.

Aristotle was, of course, more at home in the detailed study of animals than in any of the various branches of mathematics. Yet, as we saw he developed an original philosophy of mathematics that served to deflate some of the more extravagant claims made for its study. More positively there was his contribution, already noted, to the development of the concept of axiomatic-deductive proof. In the *Posterior Analytics*, as noted, he points out that the primary propositions on which deductions depend, axioms, definitions and hypotheses, must themselves be self-evident, else they would need to be themselves demonstrated and so would not be primary. With those secure, valid deductive inferences will establish not just the truth but also the certainty of the conclusions. How far any previous mathematician had progressed towards such an explicit concept is controversial. On the one hand, most of Aristotle's actual examples in that book are mathematical. On the other, not all are and he evidently considered that the model of demonstration he thereby made explicit was applicable in other domains besides the mathematical. I shall come back later to some of the difficulties of that extension.

EUCLID AND LATER DEVELOPMENTS

After Aristotle, several bifurcations in the fortunes of the mathematical sciences take place. To start, the first more or less complete systematization of most available mathematical knowledge was the achievement of Euclid, drawing no doubt heavily on the studies of his predecessors. There is no direct evidence that he knew Aristotle's *Posterior Analytics*, but the model of demonstration in his *Elements* bears a general similarity to Aristotle's. However, Euclid's primary indemonstrables differ from Aristotle's notably in that they now included postulates, especially the parallel postulate which is the fundamental assumption on which his geometry is based. This states that non-parallel straight lines meet at a point, though some later mathematicians thought that had no business among the postulates but was rather a theorem that should be proved. As is well known, the successive attempts to demonstrate it were eventually, in the seventeenth century, to lead to the development of non-Euclidean geometries.

But aside from this ambition at systematization, stunning original work in a variety of branches of mathematics was achieved by both Apollonius of Perga studying conic sections (cf. Fried and Unguru, 2001), and even more especially by Archimedes himself, not just in fairly elementary works such as *On the Sphere and Cylinder*, but in such advanced studies as *On Spirals* and *On the Quadrature of the Parabola* (Netz 2017, cf. 1999 and 2009). In Archimedes we see the power of Greek mathematical proof both in what we should call pure mathematics and in applied, that is such fields as statics and hydrostatics. Given that he is able to build on Euclid's results, he does not need to set out all his axioms. But rigorous deductive proof is everywhere the ideal. Even when as in the *Method* he is prepared to develop less strict methods in the services of heuristics, the results discovered have thereafter to be made the subject of rigorous demonstration.

Yet, to get these famous developments into proper perspective several qualifications need to be added. First, it is far from the case that every Greek mathematician was intent on producing demonstrations on that Euclidean model. Plenty of work evidenced in the writings of Hero of Alexandria in the first century CE in particular has no such ambition (Cuomo 2001). The traditions of the attack on problems of mensuration that go back to the Greeks' Near Eastern neighbors continue throughout the Mediterranean area and throughout all periods of antiquity (Høyrup 1994). An interest in mechanical devices, some of practical use such as the crane and the pulley, some just for show or entertainment, like automata, is also widely attested. Interestingly enough, although Hero makes no bid for the intellectual superiority of his investigations, he claims that they are better than traditional "philosophy" at securing peace of mind. Simplifying drastically, the view is that if you want to be free from anxiety—as the philosophers repeatedly insisted on—what you needed was not some philosophical theory about the Summum Bonum, but rather better weaponry than your opponents. That was to be achieved not by adopting some philosophical doctrine, but by the sustained study and application of mechanics (Tybjerg 2004).

DEMONSTRATION IN GALEN, PROCLUS AND IAMBLICHUS

But then the second qualification that needs to be made relates to the way in which the applicability of the axiomatic-deductive model of demonstration was exaggerated. Already in Aristotle there is a question mark as to whether or how far it was indeed relevant to most branches of the study of natural phenomena. But when we turn to the second-century CE physician and philosopher, Galen, we can see that his attempts to provide axiomatic proofs in the study of what we would call the physiology of the human body or its pathology were quite misleading (Lloyd 2006a and c). His examples of self-evident axioms include, first that nature does nothing in vain, but so far from being self-evident, this was highly controversial. A second principle was that "opposites are cures for opposites," but everything there depended on what counted as an opposite, and to have the principle come out true tended to make it vacuous. A third was that nothing happens without a cause. But even that had been denied by the Epicureans who, to secure free will, imagined that some events (the swerve) are uncaused. Of course, Galen would have protested that such a doctrine was unsustainable and deeply mistaken. But that does not alter the point that so far from being a self-evident truth, this principle was hotly contested.

Galen thus provides one example where the bid to adopt a basically mathematical model of proof in other areas of science led to considerable difficulties. Let me jump ahead now to consider an even more striking example namely Proclus' attempt to do the same in the realm of theology. Proclus was responsible for a history of mathematics which is our main source for those whose writings themselves are not extant. He repeats the common Greek belief that geometry originated in Egypt, arising from the need to cope with the problems of land measurement caused by the flooding of the Nile. But his view of the role of Euclid is that the *Elements* was geared all along to solving the problem eventually tackled in Book XIII, namely the construction of the five regular solids that Plato had drawn attention to. For Proclus Euclid is a Platonist, through and through. Then in Proclus' own *Elements of Theology* we find what purports to be a Euclidean systematization of doctrines of the divine, proceeding from indemonstrable primary propositions to the proof of such doctrines as the benevolence of God. The ideal of axiomatic-deductive demonstration, which was absent from other ancient traditions of mathematics, proved something of a mirage for some Greek ambitions to secure incontrovertibility in their worldviews.

Proclus lived in the fifth century CE, by which time revived versions of what passed as Platonism and as Pythagoreanism had had several periods of fashion at least in some quarters. At the turn of the third and fourth centuries CE, the neo-Pythagorean Iamblichus shows how far some were prepared to go. In his *On the Common Mathematical Science*, he develops an extreme version of the notion that mathematics is the key to science, invoking the Pythagoreans as ancient authorities for just such a view (cf. Sambursky 1962: 47ff.).

The Pythagoreans, he says (ch. 23), valued mathematics and applied it in many different ways to the study of the cosmos. "They considered what is possible and impossible in the structure of the universe on the basis of what is possible and impossible in mathematics, and they apprehended the celestial revolutions with their causes according to commensurate numbers." Mathematics, in fact, he goes on to say (ch. 32), enables one to understand not just the movements of the heavenly bodies, but many other natural phenomena as well:

> It is also the custom of mathematics sometimes to attack mathematically perceptible things as well, such as the four elements, by using geometry or arithmetic or harmonics, and similarly with other matters. For since mathematics is prior in nature, and is derived from principles that are prior to those of natural objects, for that reason it constructs its demonstrative syllogisms from causes that are prior . . . Thus I think we attack mathematically everything in nature and in the world of coming-to-be.

PTOLEMY

With Iamblichus we reach, as I said, an extreme where mathematics is used in the service of theurgy, that is to bend the divine to your will. But we have one final set of examples to consider to illustrate both the powers of a program of mathematical study and the difficulties its implementation faced. This will enable us to return to territory that is more familiar to us. I refer to the impressive corpus of works of Ptolemy, that spanned astronomy, harmonics, optics, and geography. In each case, he drew extensively, naturally enough, on earlier work, including detailed records of Babylonian observations that go all the way back to the first year in the reign of Nabonassar (747 BCE). At the same time as will be shown elsewhere in this volume, his own original contributions were considerable and proved to be enormously influential.

At the outset of his astronomical masterpiece, which he named the *Mathematical Composition* (*Syntaxis*) he positions mathematical astronomy in relation to both "physics" and "theology." By "physics" he understands the study of the world of change in the sublunary sphere and by "theology" the study of god, conceived as invisible and unchanging. "Mathematics," the mathematical study of the heavens that he is about to embark on, is superior, he claims, to both those other two. Both of those are matters of conjecture rather than of scientific understanding, theology because of its utter obscurity, and physics because of the instability of its subject-matter. "Mathematics" alone yields unshakeable knowledge, precisely because it does so by means of indisputable arithmetic and geometric demonstrations.

Here, too, then we find the theme we have become familiar with, namely that the great strength of the mathematical sciences lies in their capacity to deliver proof and certainty. Ptolemy does not set out his axioms in Aristotelian or Euclidean style, although he is at pains to justify his main assumptions, including that of the sphericity of the heavens, and the sphericity of the earth, as well as the doctrine that it is at rest in the center of the universe, in the opening book of the treatise. But it is thanks to its use of mathematical procedures that astronomy occupies the preeminent place it does in human intellectual disciplines.

But there is more. In the very first chapter of the work we have a text that sets out in the clearest possible terms the wider ambitions of the study of the heavens, not merely intellectual but also cosmological and moral ones. Mathematical astronomy does not just give us knowledge and understanding but leads to an appreciation of the beauty and order of the heavens. Then he has this to say about its influence on the human character:

> Of all studies this one especially would prepare humans to be perceptive of nobility both of action and of character. When the sameness, good order, proportion and freedom from arrogance of divine things are being contemplated, this study makes those who follow it lovers of this divine beauty, and instils, and as it were makes natural, the same condition in their soul. (*Syntaxis* I 1, 7.17ff)

The moral implications of the study of the heavens are given another twist when he considers the status of that other main branch of the subject, namely astrology, in his *Tetrabiblos*. The aim of astrology, like that of astronomy itself, is predictive, but its predictions relate not to the unchanging movements in the heavens, but to events on earth, and the problem here is that "every study that deals with the quality of matter is conjectural." Indeed, Ptolemy dismisses much earlier work, including the schemata that he attributes to the "Chaldaeans," as unsound and yet he holds that predictions about events on earth are possible, even if always a matter of conjecture. Moreover, this study, too, can be justified on the grounds of its contribution to well-being, in that it enables humans to face the future with calm and steadiness (*Tetrabiblos* I, ch. 2).

The program that Ptolemy outlines for mathematical astronomy is absolutely clear: it aims to provide indisputable demonstrations of the movements of the sun, moon and planets. But how far, we must ask, was such a program achieved in practice? The geometry of the construction of his models is crystal clear: it will be analyzed elsewhere in this volume. Yet, the parameters that have to be used in their application to each of the heavenly bodies studied are extremely complex. The main periods of revolution in every case, as determined by sustained observations going back to the Babylonians, do not correspond to values expressible in simple numbers. On many occasions, to allow the argument to proceed, Ptolemy allows himself approximations. This applies in particular to the value he gave to the precession of the equinoxes, which he chooses to represent as 1° in 100 years, when the data on which he based that in *Syntaxis* Book VII, ch. 2, suggest that that should have been taken as a lower value rather than the mean.

Then, to get a better fit between data and geometrical models, he has to complexify the latter. Neither a simple eccentric model, nor an epicyclic one, will do. So, they have to be combined and indeed in some cases he has to have recourse to a further complicating element, the so-called equant point. So, for most of the planets (the exception being Mercury), the planet moves on an epicyclic circle, the center of which itself moves round an eccentric circle. But the movement is uniform not with respect to the center of that eccentric circle, but in relation to the equant point, which is a point on a line from the earth through that center and equidistant from that center (see Chapter 5 for illustrations of these ideas).

Of course, this new item in the geometry was later to be greeted with derision, for it seemed quite arbitrary and to have no physical correlate. Yet, complex though the models were, they enabled remarkably accurate predictions of the movements of the planets to be made, and the *Handy Tables* that set out their cycles continued to be usable for centuries after Ptolemy's death. Yet, from time to time, Ptolemy's own sense of the difficulties he faced surfaces in the *Syntaxis* itself. Thus, when he comes to consider movements in latitude (north and south of the ecliptic), he has this to say (*Syntaxis* XIII ch. 2, H II 532.12ff. Toomer 1984: 600):

> Let no one, considering the complicated nature of our devices judge such hypotheses to be over-elaborated. For it is not appropriate to compare human [constructions] with divine, nor to form one's beliefs about such great things on the basis of very dissimilar analogies . . . Rather, one should try, as far as possible, to fit the simpler hypotheses to the heavenly movements, but if this does not succeed, [one should apply hypotheses] which do fit.

We have to assume that the heavens are simple, even though they do not seem so to us. We have, in other words, to invoke a quite different notion of "simplicity." We are evidently a far cry from the optimistic claims in the first book, that the study of the "good order" and "freedom from arrogance" of heavenly movements will enable us to achieve good order in our own souls.

CONCLUSIONS AND COMPARISONS

I have surveyed a small sample of the different understandings and uses of what the Greeks called *mathēmatikē* and how they constitute a mathematical worldview. As remarked at the outset of our study, the different views entertained on the subject are striking: it is not as if Greco-Roman culture imposed either a single conception or a unified set of practices on the subject. We, for our part, must come to terms with how their actors' categories differ from our own.

At one end of the spectrum there are plenty of instances of notable achievements in arithmetic, geometry, harmonics, statics, hydrostatics, mechanics, and astronomy that we can recognize as such. Many of these were driven by an ambition not just to discover theorems, but to demonstrate them, indeed to do so using methods that would be incontrovertible. Yet, we have also seen that much work had no such ambition, and we have seen examples where some of the claimed successes involved a fair dose of wishful thinking. Indemonstrable but self-evident primary premises were always harder to come by than the advocates of the method appreciated. Even where such axioms were not invoked, the application of geometrical or arithmetical models to physical phenomena, in astronomy, harmonics or optics, always involved a degree of idealization, where the complexities, even the messiness of the data, were glossed over (Lloyd 1987: ch. 6).

Then there are the, to us, less familiar metaphysical, moral and religious associations present in the background, sometimes in the foreground of those engaged in "mathematics." Among these we may include a preoccupation with number symbolism and especially the desire to see in mathematics not just a tool to understand phenomena, but the best way of cultivating the soul. This was a cardinal part of Plato's program, but even Aristotle, who as we have seen was so often contemptuous of the views of his fellow Greeks about mathematics, held that intellectual activity, exemplified by mathematics as well as by dialectic, is the highest activity of which humans are capable, and thus a major contributor to our well-being and happiness. Whether it is Pythagoreans swearing by the *tetraktys*, or Plato bidding us cultivate our reason, or Aristotle suggesting we should "immortalize ourselves as far as possible," or Iamblichus claiming that number is the key to all understanding, we find example after example of the grip that some vision of what "mathematics" stood for had on the Greek imagination.

As a final coda to our discussion, let me come back to some of the similarities and differences between the Greco-Roman and the Chinese experience to underline some of the difficulties of generalizing across cultures while we may recognize the commonalities that link them. Like Greek *mathēmatikē*, what the Chinese included under their rubric of *shu* or *suan shu* (the study of numbers) covers far more than our "mathematics." Depending on context, *shu* can mean "fate," "destiny," "art," "deliberation," and even "reprimanding" as well as "counting" or "number" (Ho 1991). The books devoted to *shu shu* (counting and methods) in the imperial library related to a number of studies that included the patterns in the heavens and calendar studies. Then another area where the ancient Chinese, like the Greeks, brought mathematics to bear to illuminate aspects of physical phenomena was harmonics, the study of the major concords and the twelve-tone scale.

As in Greece, so, too, in China, mathematics comprised not just practical skills of use in everyday life, and in the Chinese case in the context of the administration of a vast empire, but also was a prime example of methods that could be used to increase knowledge. The Chinese did not develop a notion and practice of axiomatic-deductive demonstration, but as the text I set out in the Annex shows, they could be impressed by the way a mathematical approach to a problem could yield understanding. But this depended not on some appreciation of the success in attaining incontrovertibility, but rather on the practitioner mastering the techniques to be used. The master does not overwhelm the pupil with proof, but encourages him or her to internalize those techniques, a matter of acquiring the *Dao*, no mere intellectual method, rather the mastery of a practice.

So, it is then that both these ancient civilizations saw their distinctive views of mathematical study as no mere intellectual activity aiming at increasing knowledge, but as an essential part of self-fulfillment. The further exploration of the actual applications of that concept in China will be the subject of detailed study in Volume 2. So far as the Greco-Roman world is concerned, we have seen the variety of manifestations of different ambitions for the subject, the disputes on what the subject consisted of, what its objects were, how it was to be justified, and how it related to the good life, indeed to the highest aspirations of which humans are capable.

ANNEX

The aspirations of "mathematics" in the Chinese text *Zhoubi suanjing* compiled at the turn of the millennium, from Cullen (1996: 176–7).

Long ago, Rong Fang asked Chen Zi, "Master, I have recently heard something about your Way. Is it really true that your Way is able to comprehend the height and size of the sun, the [area] illuminated by its radiance, the amount of its daily motion, the figures for its greatest and least distances, the extent of human vision, the limits of the four poles, the lodges into which the stars are ordered and the length and breadth of heaven and earth?"

"It is true," said Chen Zi.

Rong Fang asked: "Although I am not intelligent, Master, I would like you to favour me with an explanation. Can someone like me be taught the Way?"

Chen Zi replied: "Yes, all these things can be attained to by mathematics. Your ability in mathematics is sufficient to understand such matters if you sincerely give reiterated thought to them."

At this, Rong Fang returned home to think, but after several days he had been unable to understand, and going back to Chen Zi he asked: "I have thought about it without being able to understand. May I venture to enquire further?"

Chen Zi replied: "You thought about it but not to [the point of] maturity. This means that you have not been able to grasp the method of surveying distances and rising to the heights, and so in mathematics you are unable to extend categories . . . It is the ability to distinguish categories in order to unite categories which is the substance of how the worthy one's scholarly patrimony is pure and of how he applies himself to the practice of understanding . . . So if you cannot apply yourself to the practice of mathematics, why should I confuse you with the Way?"

Eventually, Chen Zi relents and explains the problems to Rong Fang, but it is notable that the latter's progress is linked to his character and to his inability to cultivate his patrimony, rather than to a merely intellectual grasp of an impersonal method that will deliver certainty.

[illegible]

CHAPTER FIVE

Describing and Understanding the World

Counting, Measuring, and Mathematical Modeling in Babylonian and Greco-Roman Worlds

FRANCESCA ROCHBERG AND J. LENNART BERGGREN

The previous chapter, "Mathematics and Worldviews," showed how ancient conceptions of mathematics ran hand in hand with conceptions of the world—indeed, in the case of the Pythagoreans, a conception of the one *was* a conception of the other. In this chapter, we will look at how mathematics as a body of knowledge was *used* to understand the world, particularly the heavens. But, as this whole book has tried to make clear, mathematics is not just one body of knowledge: its methods and emphases are culturally determined. Accordingly, the role of mathematics in understanding the world, too, is culturally determined, and in ancient times, even where there was interchange, the mathematical astronomy of the Babylonians and the Greeks, for example, bore a distinct cultural stamp.

Thus, numeracy in cuneiform culture which allowed Babylonian scribes to manage a sophisticated administrative bureaucracy and a complex economy, also shaped the first empirical and quantitative astronomy—a rich astronomy including relations for the cyclical movements of the sun, moon, and planets, the introduction of the system of degrees for recording zodiacal positions, and records of observations going back to the early eighth century. All of these were ingredients for astronomical modeling by linear mathematical methods which were imported to the Greek world during the Hellenistic period (after the 2nd century BCE) and adapted for both astronomy and astrology.

Greek astronomy, however, focused on predicting positions of heavenly bodies at any given time instead of the Babylonians' interest in synodic moments. Most importantly the Greek astronomers and mathematicians expanded their methods to include geometry and trigonometry. This Greek development in geometrical methods and modeling enabled new mathematical treatments of the dimensions of the cosmos. Some of the same individuals who worked in this geometric tradition also applied astronomical data and geometrical methods to create planar maps of the known world that gave a reasonable approximation to the earth's curved surface and were, even in the fifteenth century, the best basis for scientific geography.

Let us now consider the mathematical approaches to the world developed by these two cultures, the cuneiform culture of the ancient Babylonians and the culture of Greek and Rome.

MATHEMATICS AS A SUBCULTURE IN ANCIENT IRAQ

In his *Mathematics as a Cultural System* (1981: 13), Raymond Wilder found that the development of mathematics as "a subculture of the general culture" did not occur until relatively later in history than in the ancient Middle East. "In Babylon and Egypt," he said, "mathematics had evidently not gained a status that would warrant calling it a 'subculture'; it was only a *cultural element*" (emphasis in the original). By "subculture," Wilder was pointing to mathematics as it is done in the modern world, an elevated scientific field practiced by specialists. But more is known about numbers and mathematics in cuneiform culture, and antiquity as a whole, today. In discussing the coefficient lists, Eleanor Robson (1999: 14) raised the question as to whether these were for "working bureaucrats" or "the display of scribal virtuosity," implying an intellectual rather than solely practical purpose for these products of knowledge, which in turn could imply the existence of scholars specialized in mathematics. The evidence from a group of Late Babylonian factorization tables is unequivocal. If there was any doubt about a scholarly mathematical culture, these texts certainly remove it (Ossendrijver 2014: 151). Speaking about the factorization tables, Mathieu Ossendrijver noted that the numbers written on these tablets are "the longest numbers found in a cuneiform text and probably in all antiquity" (Ossendrijver 2014: 150). For example, Text A gives the value of 9^{46} as 16.34.3⌜9⌝.[52.40.21.26.52.57.35.56.49.50.37.38.58.13.38.4.44.57.15.3.37.21] (where periods separate sexagesimal digits, square brackets indicate breaks in the tablet, and half-brackets a half-broken sign). Ossendrijver said, "By exemplifying unparalleled, supra-utilitarian numerical skills and, perhaps, a notion of numerical proof and an interest in number-theoretical regularities, factorization tables shed significant new light on Late Babylonian mathematics, for which they imply a greater autonomy from mathematical astronomy than previously thought" (2014: 151). This would attest to a subculture of mathematical specialization, or at least a field in which such "supra-utilitarian numerical skills" were defining. This evidence from the Late Babylonian period, or Hellenistic Babylonia, represents the end of a long tradition, beginning with the beginnings of writing itself.

The importance of numbers and counting to the cuneiform (and proto-cuneiform) writing system was a function of the original purpose of script to record the economic data used by the early Sumerian temple administrative bureaucracies of southern Mesopotamia. Thus numbers, number notation, counting in various base systems (decimal [10], sexagesimal [60], bisexagesimal [120], a system for area measure [GAN_2], a system for measuring time [U_4]), and metrology were present in ancient Iraq from the very beginning of written texts. From this utilitarian foundation in the fourth millennium BCE, mathematics in the cuneiform world developed into a scholarly specialization of its own, not justified by administrative or even predictive astronomical needs, but as its own justification for exploration of the properties of integers, or number theory. However, the exploration and use of numerical relationships for describing certain phenomena or parts of the world, as well as for calculating cyclical heavenly phenomena, was a long-standing use of mathematics in cuneiform culture.

The only cuneiform number notation system to survive into the third millennium was the sexagesimal or base 60. For mathematical and astronomical texts, the cumulative-positional sexagesimal system was the exclusive notation system. Therefore, the numbers in the following will be written in the modern convention for sexagesimal numbers, where the comma (,) separates sexagesimal places, e.g., 1,12 =72, and the semicolon (;) separates integers from fractions, e.g., 1;12 = 1+12/60 or 1 and 1/5.

THE ROLE OF NUMBERS AND MEASURES IN IMAGINING AND DESCRIBING THE WORLD

From the Old Babylonian period (*c.* 1800–1600 BCE), as a consequence of the growing variety of practical purposes to which mathematics was applied, such as the calculation of the areas of fields, volume of granaries and storerooms, or the dimension of canal trenches, mathematical relationships came to be of interest, as evidenced in the two primary mathematical text types, the list, or table text and the problem text. Problem texts cover subjects sometimes related to real situations, sometimes rather as exemplifications of procedures for solutions. Some are terse statements of a problem, for example:

> The side of a square is 1 UŠ. I made a border each side and I drew a circle (see Diagram 5.1). What is its area? (BM 15285, redrawn after Robson 1999: 209)

Other problem texts offer worked examples, such as:

> [A heap. The circumference is 0;30, the height 1.] In 1 cubit, [what is the slope? You:] Take the reciprocal of 1, the height. You will see 1. [Multiply 1 by 0;30, the circumference.] You will see 0;30. Break 0;30 in half. You will see [0;15]. The slope is 0;15 in 1 [cubit. This is] the method. (BM 96954+, see Robson 1999: 224)

The table texts list numerical relationships such as reciprocals, Pythagorean triples, or various coefficients termed IGI.GUB/*igigubbûm*—sometimes translated as constants, key numbers, or constant factors.

Coefficients were used for calculating variable quantities in many contexts. They are of two sorts, those concerned with strictly mathematical relationships (for circles, or radii, diameters) and those with culturally conventional relationships (the daily work rate for making bricks, brick density, storage). As an example from the realm of administrative accounting, coefficients were established for calculating the volume of storage of a variety of materials. The

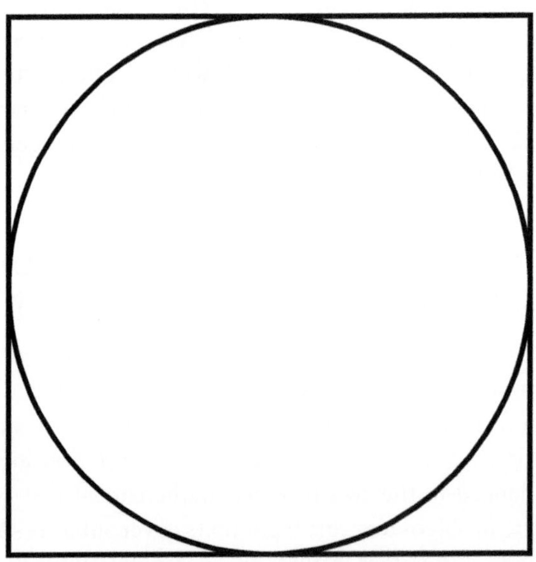

DIAGRAM 5.1. BM 15285: The area of a circle in a unit square.

following is a compilation from several coefficient lists of various storage coefficients (Robson 1999: 122 §7.4.1):

A storehouse: 0;3,36 is its coefficient.
0;3,36, of the storage of straw.
15, of the storage of water.
50, a cargo-boat of reeds.

In the case of astronomy, the use of coefficients quickly became an important mathematical tool used in the description and modeling of celestial phenomena (see below). Coefficients used for astronomical calculation of the length of daylight and the duration of lunar visibility were included in the lists devoted to coefficients, referred to before. The following (discussed further below) appear in such coefficient lists (Robson 1999: 129 §8.2.1):

12, the water-clock.
3, 45, the visibility (of the moon).
40, to change day into night.

A group of ideal units of time reckoning was devised for accounting purposes, entered into early Babylonian astronomy as well as the tradition of *Enūma Anu Enlil* omens, to remain in the cuneiform scholastic tradition for two millennia, for example the *bēru* "double-hour" and UŠ "degree," where 12 *bēru* = 1 day, 1 day = 360 UŠ, and 1 UŠ = 4 minutes (of time). Of the local Sumerian calendars in the Ur III period, where real month lengths varied, calendar months in the city of Nippur were standardized and later taken over as the month names of the ideal calendar (12 30-day months = 1 ideal year) common to the scholarly traditions of astrology (celestial and natal divination) and astronomy (MUL.APIN, Astrolabes) prior to *c.* 500 BCE and even later in some cases. The ideal calendar survived in Late Babylonian (4th to 1st century BCE) astrological texts, discussed below.

Weight measures were also incorporated into texts dealing, for example, with the length of daylight, and may have reflected the use of a water clock where the weight of water measured in *mina* indicated a length of time: 1 ½ MA.NA KI.LÁ 9 UŠ *ina qaqqari* "1 ½ minas weight (of water in the water clock) is (equal to) 9 degrees linear distance" (see CAD sub verbo ***manû*** A meaning 3 b "mina" where 1 mina = 1/60 of a talent). In early Babylonian astronomical texts where units are sometimes not expressed, either UŠ or mina can be interpreted, where 1,0 (60) UŠ = 1 mina. For observational purposes, units familiar from the texts of land surveyors, such as the ***ammatu*** cubit or "forearm" and ***ubānu*** finger, are also found in astronomical contexts where observational distance between celestial bodies is of interest. Discussion of measurement in astronomy is found in Steele (2007).

What is salient in the present context is that the description of celestial phenomena as known by ancient Mesopotamian scholar-scribes took the form of mathematical models. The relationship between scientific modeling and the empirical world is an important one today and, indeed, it was one of the principal cognitive routes to representing parts of the world in ancient Iraq as well. The elements of a model need not replicate or simulate the object of the model in a physical way, but if there is the fundamental relationship of reference, or denotation, between model and world, then the attempt to model phenomena of the world is a *form* of world description.

Put another way, the process of modeling, regardless of the physical relation between the actual model and the world (the spherical Greek planetary models are a good example, see further below), aims to denote or refer to some phenomenon, or phenomena, of the world that makes sense in terms of some overall world picture. Using the Greek cinematic astronomical models as a counter example to those from the cuneiform world, the Greek models made sense in terms of the geocentric spherical world picture. The Babylonian models, on the other hand, do not reflect that image, but derive from a conception of the world constructed of numbers, measures, and numerical relationships. This denotative relationship between the mathematical models of certain astronomical phenomena by cuneiform scribes and the world they imagined as surrounding them, is what establishes the connection between ancient Babylonian astronomical models, or modeling, and the history of science in general.

ASTRONOMICAL SCHEMES AND MATHEMATICAL MODEL-MAKING, (1800–250 BCE)

By the early second millennium BCE, evidence that Babylonian scribe-scholars found certain cyclical phenomena to be amenable to mathematical description is found in the earliest astronomical texts, and the phenomenon first to be so treated was the variation in the length of daylight through an ideal year of twelve 30-day months. The continuous increase and decrease of the duration of daylight, repeated every 12 cycles stands as the first attested object of a mathematical model of a periodic phenomenon, and it took the form of a "piecewise linear zigzag function." Values of such a function increase or decrease by a constant increment or decrement between two extrema, a maximum and a minimum, repeating at a periodic rate (see the schematic presentation in Diagram 5.2).

The earliest evidence for the quantitative model for daylight length is found in an Old Babylonian text (BM 17175+, see Hunger and Pingree 1989:

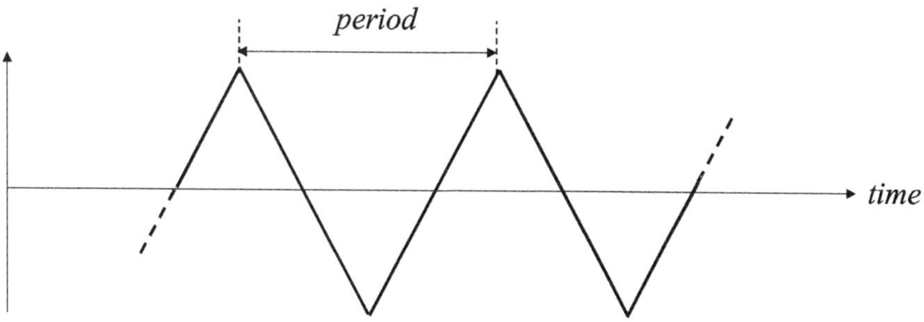

DIAGRAM 5.2. Piecewise linear zigzag function.

appendix, 163–4). In four sections, one for each schematic season, the text gives the model as follows:

> [On the 15th of Addaru, 3 (we can supply minas, or 3,0 UŠ) are a wa]tch of the day, 3 (minas, or 3,0 UŠ) are a watch of night; [Day and night] are equal. [From the 15th of Addaru to] the 15th of Simanu is 3 months. [On the 15th of Simanu, the night] transfers 1 (mina, or 1,0 UŠ) of the watch to the day. [. . . 4 (minas, or 4,0 UŠ) is the wa]tch of the day, 2 (minas, or 2,0 UŠ) is the watch of the night. (square brackets indicate restored passages broken away from the clay tablet)

This model of the ideal year assigned the equinoxes and solstices to the midpoints, or 15th day, of months XII, III, VI, and IX. For each schematic season, or quadrant in the ideal year, the length of daylight shifted by 1 unit. Therefore, from vernal equinox to summer solstice, the length of day increased by "1," from summer solstice to autumnal equinox, daylight decreased by "1," and so on, producing a model for the change in the length of daylight in which the ratio of longest to shortest day length was 2:1. Thus: 3 (VE)+1 = 4 (SS) - 1 = 3 (AE) - 1 = 2 (WS)+1 = 3 (VE). This scheme is not practical for the geographical latitudes of Mesopotamia, but it is the simplest, indeed most elegant, way to model the experience of increasing and decreasing durations of daylight around two extremes (summer and winter solstices), provided the model is constructed on the ideal year. The mean value was expressed in sexagesimal notation as the number 3, i.e., 180 (3 x 60), and represented one-half of the circle of the day (360 degrees) when daylight and night were of equal length (180 degrees). In order that the length of daylight increase by "1" from VE to SS, and decrease by "1" from SS to AE, and AE to WS, and then increase again by "1" from WS to VE and thereby complete one cycle, the monthly constant difference (or coefficient) for this model had to be 20, as illustrated in Table 5.1.

TABLE 5.1. Astrolabes' scheme for variation in daylight length

Month	*C (in mana)**	*C in UŠ**	*= Hours*	*Cardinal Points*
XII	3;0	3	= 12hr	Vernal Equinox
I	3;20	3,20	= 13hr 20'	
II	3;40	3,40	= 14hr 40'	
III	4;0	4	= 16hr	Summer Solstice
IV	3;40	3,40	= 14hr 40'	
V	3;20	3,20	= 13hr 20'	
VI	3;0	3	= 12hr	Autumnal Equinox
VII	2;40	2,40	= 10hr 40'	
VIII	2;20	2,20	= 9hr 20'	
IX	2;0	2	= 8hr	Winter Solstice
X	2;20	2,20	= 9hr 20'	
XI	2;40	2,40	= 10hr 40'	

Note: *The relation between these measures is 1 *mana* = 1,0 UŠ. Mana was a unit of weight, for measuring water into the water clock.1 UŠ = 4 time degrees.

A group of texts modern scholars termed "Astrolabes"—not to be confused with the metal astronomical calculating instruments of the same name—were cuneiform texts with a circular or list form showing three groups of stars in their heliacal risings, i.e., first appearances after invisibility due to conjunction with the sun. These stars, three for each month, are listed in columns or arranged in circular formation, month by month together with numerical values for the length of day in those months. This group of texts provided the full complement of numerical values that made up the model attested in the Old Babylonian Example cited above. In Table 5.1, "C" designates the value for length of daylight, taken as constant for each day of one of the twelve months. In other words, the daily variation was not taken into account, only the monthly. The mean value of the table is 3 (= 3,0), representing the length of daylight (or night) at the equinoxes.

A comprehensive treatment of fixed stars, planets, and celestial phenomena from the early first millennium BCE is the two-tablet series entitled MUL.APIN, "The Plow Star" (Hunger and Steele 2019). It, too, includes reference to the model for the change in the length of daytime consistent with the Old Babylonian version just described.

The calendrical underpinnings of MUL.APIN are schematic (the year of 12 30-day months), as are its catalogues of stellar risings and settings, and intercalation schemes. The text states, for example: "On the 15th of Month I, a nighttime watch is 3 minas; 12 UŠ the (coefficient for the daily retardation of the) rising of the moon" (Hunger and Steele 2019: 95, Tablet II ii 44). The text further spells out the lengths of night for each ideal month but also included the value for the visibilities of the moon, whether from rising or before setting. MUL.APIN's interest in the night lengths and visibilities of the moon is followed by a short passage explaining the calculations for the duration of lunar visibility using a "difference" coefficient (*nappaltu*), e.g., "multiply 40 NINDA, the 'difference' of daylight and night (*nappalti ūmi u mūši*), by 4 and you will find 2,40, the 'difference' of the visibility (*nappalti tāmarti*) of the moon" (MUL.APIN II iii 15, see Hunger and Steele 2019: 101).

Also focused on describing the variation in the duration of lunar visibility using a mathematical scheme is *Enūma Anu Enlil* Tablet 14 (Al-Rawi and George 1991–2). It provides a tabulated scheme for the length of visibility of the moon each night for the 30 days of the two equinoctial months (when day and night are of equal length as the sun crosses the equator in spring and autumn). The interest in duration of lunar visibility is probably tied to the ominous nature, whether for bad or good portent, of the moon when visible. This table, and a second table in the same tablet that gives a supplementary tabulation of lunar visibility coefficients allowing calculation of lunar visibility in other months of the year, is underpinned by the arithmetical scheme for the variation in daylight seen in the Astrolabes and MUL.APIN, just discussed.

The duration of lunar visibility was, of course, related to the length of night, and the value given as the IGI.DU$_8$.A = *tāmartu* "visibility" of the moon is calculated as 1/15th of the length of night. For example, in Month I day 1, day = 3 and 1/6 and night = 2 and 5/6 (3 here is in fact 3,0 or 180). On this day, the IGI.DU$_8$.A of the moon is given as 11;20, the result of dividing the length of night by 15. For an equinoctial day, e.g., Month I, day 15, the IGI.DU$_8$.A of the moon is given as 12. Night length at the equinox = 3,0 (moon rises at sunset and rises at sunrise and is visible the entire night, for 180°). These values were results of modeling, not measuring, the variation in length of daylight and the underlying structure was the ideal year, 12 30-day months or 360 days. The numerical mean value in this ideal scheme, the value 3 (3,0 = 180), as the representation of one-half of the circle of the day, was not simply a derived mean value from the schemes for daylight length and duration of lunar visibility, but played a determining role in the construction of those schemes. Similarly, for use in later Babylonian mathematical astronomy, mean values were derived for units of time such as the mean synodic month and the mean *tithi* (1/30 of a mean synodic month). Mean values can also be derived from Babylonian tables for computing synodic arc ($\Delta\lambda$), or progress in longitude made by the planet or moon per synodic phenomenon, and synodic time ($\Delta\tau$), or the time required for the body to complete a synodic cycle between successive phenomena of the same kind (e.g., first visibility to next first visibility). Such mean values determined the structures of lunar and planetary theory.

The Late Babylonian astronomy, from approximately 480 to 45 BCE, is typified by the application of methods, or "systems," of calculating the positions and dates of the synodic phenomena of the moon and planets. These calculation systems entailed an understanding that heliacal appearances were the result of the distance (elongation) of the planet or moon from the sun, which is to say, the planet's direction in the sky in relation to the sun, e.g., at opposition, 180°. Therefore, an intimate connection was found between synodic arc ($\Delta\lambda$) and synodic time ($\Delta\tau$).

Late Babylonian mathematical astronomy is represented in table texts, numbering approximately 450 tablets, as well as procedure texts, which provide instructions for how to carry out the computations (not unlike the mathematical problem texts giving worked examples, mentioned before). Underlying the computational methods of the tables was the recognition of period relations such as the nineteen-year cycle (see below for the Metonic cycle), or the Saros, or the synodic periods of the five classical planets. In these "ephemerides" (true ephemerides with daily positions are a rarity, e.g., ACT 654+655 for Jupiter, and 310 for Mercury), two types of recursive mathematical steps (algorithms) were applied for calculation of the synodic arc. In the planetary System A, for example, the synodic arc $\Delta\lambda$ and synodic time $\Delta\tau$ are stretch-wise constant in different "zones" of the ecliptic, with various rates of modification from one zone to the next, producing what we call a step function in modern mathematics. In System B, on the other hand, the differences $\Delta\lambda$ (or $\Delta\tau$) form a zigzag function, the same mathematical structure seen in the calculation of the daylight scheme mentioned above, except the underlying period (in planetary ephemerides) is now based upon the relation between the number of synodic occurrences, sidereal rotations of the planet, and the number of Babylonian (sidereal) years. The construction of both systems took place early in the Seleucid era, with chronological priority going to System A (Ossendrijver 2012: 116).

In the construction of mathematical models to generate the positions and dates of the table text, as in earlier astronomical modeling, the use of coefficients is more complex than in the Astrolabes or *Enūma Anu Enlil* 14. In the calculation of the synodic phenomena of the planet Jupiter, for instance, according to System A, an interpolation coefficient is used when crossing the boundaries of a 2-zone model of planetary progress in longitude $\Delta\lambda$. According to such a model, the body progresses at one rate (so and so many degrees per time period) in one zone of the model, at a different rate in the other zone.

A procedure text (ACT 821, see Ossendrijver 2012: 281 Text Nr. 28) says:

> [From 25° Gemini until 30° Sc]orpius you add 30. From 30° Scorpius until 25° Gemini you add 36. (The amount) by which [it exceeds 25° Gemini you multiply by 0;50, add to 25° Gemini, and put down. (The amount) by which it exceeds 30° Scorpius] you multiply [by 1;1]2, you add to 30° Scorpius, and put down.

The procedure text shows that longitudes on the ecliptic, or zodiacal circle, expressed as 1°-30° of a zodiacal sign are calculated according to the 2-zone model. In order for the model to account for the change in "velocity" from one zone to another, interpolation coefficients, the numbers 0;50 (5/6) and 1;12 (6/5) given in the procedure text, are derived from the ratios of the rates of progress 30 and 36, viz., 30/36 = 5/6 or 0;50 and 36/30 = 6/5 or 1;12, and the interpolation coefficients are therefore reciprocals 6/5 × 5/6 = 1. As the planet in one of its synodic phenomena crosses from, say, the slow zone (30° per synodic period) into the next, or fast zone (36° per synodic period), the appropriate coefficient comes from the ratio of progress in the fast zone to that in the slow, or 36/30 = 6/5 or 1;12. By this method, the longitude of a synodic appearance of a planet can be generated from the model.

The relation of zodiacal positions, or planetary longitudes, to synodic times, or the periods required for the planets to progress in their cyclical appearances, was possible because of the use of the unit 1/30th of a synodic month. Both zodiacal signs and synodic months, therefore, were subdivided into 30 parts each. The progress of the sun around the zodiac, therefore, could be taken as 1°/day, where "day" meant 1/30th of a synodic month. Indeed, this unit was called "day" (UD) in the Babylonian texts.

IMAGINING THE WORLD WITH MEASUREMENT AND NUMERICAL RELATIONSHIPS

Apart from astronomical texts as such, measurements and means in parts of the world emerge in scholarly cuneiform texts as having a certain thematic importance in the way the world was imagined. Divination employed the sense of a norm by means of which a phenomenon could be judged to be normal or abnormal, thus portending good or ill. That is, what was normal could be gauged in terms either of a "mean of measurements" or an ideal, a "positive principle of evaluation," where that ideal is determined by the divine scheme of things (Brown, Fermor, and Walker 1999–2000: 137–8, for this idea in relation to the scheme for the length of daylight).

The idea of order by measurement, and what is ideal, was also transmitted in the contexts of cosmogonic mythology and astrology. In the corpus of mythological literary texts, measurement can be found in a number of texts describing parts of the world. In the creation poem *Enūma Eliš*, Marduk measures the size of the world regions of *Apsû*, *Ešarra*, and *Ešgallu* in order to make the world well proportioned. The importance of the idea of measured counterparts, equivalents, is also clear in *Enūma Eliš* IV 141–6 (Lambert 2013: 95 lines 141–6):

He (Marduk) crossed the heavens, surveyed the sky, and he adjusted them as

> the equivalent of the *Apsû*, Nudimmud's abode. Bēl measured the shape of the *Apsû* and set up *Ešarra*, a replica of *Ešgalla*. In *Ešgalla*, *Ešarra* which he had built, and the heavens, he settled in their shrines (the gods) Anu, Enlil, and Ea.

Marduk's own "house," the temple Esagil, in *Enūma Eliš* VI 62, is said to be "the equivalent" or "counterpart" to the great abyss, the *Apsû*, the watery subterranean region where Ea dwelled: "They raised the peak of Esagil, a replica of the *Apsû*" using the term *miḫirtu* "equivalent, counterpart." In the account of the rebuilding of Marduk's temple in Babylon by the Assyrian king Esarhaddon, deliberate reference to the Babylonian *Enūma Eliš* is made, repeating that Marduk's temple Esagil is the counterpart of the *Apsû* (*maṭṭalāt apsî*) and the equivalent, or likeness (*tamšīlu*) of *Ešarra* (Leichty 2011: 198 Esarhaddon 104 iii 41b–iv 1, and 206 Esarhaddon 105 iv 37b–v 15). The king conveys the care he took in restoring this temple by emphasizing the precise measurement of its dimensions:

> [In] a favorable month, on a propitious day, I laid its (Esagil's) foundation platform over its previous foundations (and) in exact accordance with its earlier plan I did not diminish (it) by one cubit nor increase (it) by half a cubit. I built and completed Esagil, the palace of the gods, an image (*maṭlāt*) of the Apsû, a replica (*tamšīl*) of *Ešarra*, a likeness (*meḫret*) of the abode of the god Ea, (and) a replica of (the square of the constellation) The Field; I had (Esagil) ingen[iously] built (and) I laid out (its) square (*mitḫartu*).

Another kind of measurement occurs in scholastic and literary texts, for example, in the ritual explanatory text KAR 307 (Livingstone 1989: 99–105, Nr. 39), which contains a curious passage that may be relevant here. Following its explanation of the Tigris and Euphrates as the right and left eyes of Tiamat (KAR 307 rev. 3), the text says:

> 40 *bēru*[1] is the disc/circumference (variously read *siḫirtu*, or simply HAB-*rat*) of the sun (Šamaš); 60 *bēru* is the disc/circumference of the moon (Sin).

Numerical relationships also took on considerable importance in Late Babylonian astrology, i.e., natal divination by reference to the planets in the zodiac. In a number of tablets dating to the pre-Seleucid period, or between the sixth and fourth centuries, one method of constructing numerical relationships was regularized to the extent that it appears in a group of table texts now referred to as "calendar texts." Lis Brack-Bernsen and John Steele (2004: 118) used the termed "mathemagics" for these texts' characteristic manner of constructing a scheme that had no intrinsic astronomical meaning, but enhanced the astrological value of a date or a position in the zodiac. The following is a group of lines from a calendar text where the columns represent zodiacal sign, degree, month, and day (in the ideal year):

10	7	1	1
7	14	1	2
4	21	1	3
1	28	1	4
11	5	1	5

(BM 96258 obv. 1–5, see Brack-Bernsen and Steele 2004: 96)

The numbers in col. I represent zodiacal signs (10 is Capricorn, for example) and decrease each line by 3 signs (modulo 12). The numbers in col. II represent degrees and increase each line by 7 (modulo 30). It can be seen that the "distance" from one zodiacal position from line to line is 277 degrees: e.g., Capricorn 7 to Libra 14 = 277 degrees; Libra 14 to Cancer 21 = 277 degrees, and so on. The numbers in cols. III and IV can be interpreted as months and days in the ideal calendar where a month always has 30 days. Thus, a correlation can be made between a position in the zodiac, such as Capricorn 7° (in line 1), and a "date" (Mo I day 1) in the ideal year, or vice versa. The equivalence of the 12 30-degree zodiacal signs with the 12 30-day months in the ideal year was a basic device in Babylonian astrological texts that enabled such "calculations," not based on actual but rather "ideal" astronomical relationships, to provide further positions for astrological interpretation.

Another kind of numerical relationship used in astrological schemata found its way into the manner in which planetary positions in the zodiac at the moment of birth could have in relation to one another. What would later, in Greco-Latin astrology, be termed the "aspects" (*aspectus*) of one planet to another, such as trine (120° separation), opposition (180° separation), quartile (90° separation), and sextile (60° separation) seem to be a Greek elaboration on an original Babylonian idea about the trine relationship, attested in a Late Babylonian eclipse omen text, as in the following example (BM 36746 obv. 5'–6' and 10'–11 ', see Rochberg-Halton 1984: 136):

> If the moon is eclipsed in Leo and finishes the watch (of night) and the north wind blows; Jupiter does not stand (in) the eclipse; Saturn and Mars stand in Aries or in Sagitarrius . . .
>
> If the moon is eclipsed in Virgo and [finishes the watch (of night)] and the south and east winds blow; Venus is not visible in its eclipse . . . Saturn and Mars . . . in Taurus or in Capricorn are visible . . .

This excerpt shows that a relationship was schematized for the position of the eclipsed moon in the case where the benefic planets, Jupiter or Venus, are not visible in the moon's own sign, but the positions of the malefic planets,

Saturn and Mars, are located in signs 120° from that of the eclipsed moon. In the quoted excerpt, the first configuration has the eclipsed moon in Leo, the malefics in Aries and Sagitarrius, thus forming a "trigon" (τρίγωνον "triangle"), hence the "trine" aspect where each sign is 120° apart. The second example has the eclipsed moon in Virgo with the malefics in trine aspect in their positions in Taurus and Capricorn, respectively.

In view of the awareness of and interest in numerical relationships, and the function of measurement in the exploration of what the world is and how to describe it, it would seem that number, number relations, and mathematics were an integrated part of cuneiform culture on more than one level, from the practical or utilitarian to the level of rarified scholarly specialization. The cultural history of mathematics in the cuneiform world shows that numeracy, the concept of number, and mathematics itself were not only part of practical life and empirical science but also belonged to a more general consideration or conception of what the world was and how it was structured. One could add that the critical elements of matter and mechanics, which together with mathematization, created modern science, were not reflected in the cuneiform corpus, but that the mathematization of astronomical phenomena in Babylonia had a decisive impact on the Greek development of astronomy and physics and a new natural philosophy.

MATHEMATICS AND ITS DIVISIONS IN CLASSICAL GREECE

The beginnings of mathematics in ancient Greece may be traced back to the Pythagoreans in the sixth century BCE, for whom the understanding of number leads to the understanding of the universe. By the time of Archytas of Tarentum (early 4th *c.* BCE), a friend of Plato, the adepts in the study of numbers had the distinctive name of "mathematicians" (from *mathēma*—lesson, something learned), whose study included "the speeds of the stars, their risings and settings and spherics" (Heath 1921: vol. 1, 11), as well as geometry, arithmetic, and music, subjects Archytas called "kindred."

What astronomy meant in Plato's circle we learn from his advice to astronomers in his *Republic*: proceed by problems, as geometers do, and "leave the starry heavens alone," since they are only imperfectly reflected in our world. Study of motions of, say, planets provides hints of the Creator's intentions, but not more than hints.

As for music, it made good use of arithmetic to describe and understand harmony. Arithmetic was not restricted to the practical art of what the Greeks called "logistic," or calculating but included the theoretical study of numbers (including what we would call numerology) and their relations to each other in the theories of ratio and proportion. Such relations were used to explain harmonious and discordant sounds in treatises from the time of Archytas to that of the famous astronomer Ptolemy of Alexandria (2nd *c.* CE).

Ratios of small whole numbers also found their realization in architecture. For example, when the architects Ictinus and Callicrates built the Parthenon in Athens (447–432) they used the ratio of the two smallest square numbers, 9:4, to determine the basic form of the whole building (Figure 5.1). So, the ratios making up the proportion 81:36:16 are equal to the ratios of length : breadth : height in the Parthenon, and the ratios in this proportion, 81:36 and 36:16, are equal to 9:4 (Artmann 1999: 121–3).

By the time of Aristotle (384–322), two generations after Plato, the list of mathematical disciplines had grown to include, for example, optics, engineering and metrology. And the theoretical view of astronomy and other sciences had changed. Thus, in his *Physics* Aristotle described astronomy—along with optics, and music—as "more physical than mathematical." And Aristotle recognized two parts of geometry as well: the theoretical, which we know from Euclid's *Elements*, and the practical which we know from Heron's *Metrika* (20–80 CE), which recorded the practice of Greek surveyors in measuring lengths and areas. Similar works formed the basis for Latin texts instructing the Roman surveyors in these topics (Dilke 1987b: 29–34).

Geminos (*fl.* 90–35), in the classification of the mathematical sciences in his *Philiokalia*, treated astronomy as a study of "things known through our senses," along with music, optics, mechanics, geodesy and calculation.

FIGURE 5.1. The Parthenon in Athens. Photo by Art Kowalsky / Alamy Stock Photo.

To this overview of the changing emphases on the ideal and the empirical could be added other examples of the uses of mathematics in ancient Greece—for example, Heron's use of stereometry in mechanics and Archimedes' (*fl.* 287–212) use of geometry in the study of centers of gravity. But we now turn to the particular mathematical sciences of astronomy and geography.

ASTRONOMY: THE EARLY STAGES

The rhythms of what was predominately rural life in Greece were guided by experience based on observations of *ta meteōra* ("the things in the air," whether heavenly bodies or rainbows or thunderstorms). As did the Babylonians before them, the earliest Greeks used numbers to model the phenomena and these numbers provided workable values for the lengths of the cycles associated with the sun, moon, and stars to establish a calendar suitable for astronomy.

One finds early traces of such calendar-making in Hesiod of Boeotia's (*c.* 700?) *Works and Days*, which contains a guide to the agricultural activities of the seasons of the year based on solstices, equinoxes and heliacal risings and settings of certain bright stars or constellations. In this context, "risings" and "settings" of stars refer not to those that occur every day and night, but to what were known as their "phases" (see the section titled, "Astronomical Schemes and Mathematical Model-Making (1800–250 BCE)," above).

The earliest mathematical treatment of the phases in Greek antiquity was that of Autolycus of Pitane (*c.* 360–290). Although he did not have the mathematics to compute the times of the true phases, he did explain and prove some basic facts about phases, and his book was studied as an introduction to the topic in ancient Greece, medieval Islam, and medieval Europe.

In the course of time, heliacal risings and settings became, quite literally, "written in stone" on *parapēgmata*, stone tablets with holes drilled for a movable peg for each day of the year (Figure 5.2), beside which information about the phases of prominent stars or parts of constellations on that day would be written, along with prognostications about the wind and weather (Lehoux 2007). Similar texts, MULAPIN for one, have been found in Babylon, and such "farmers almanacs" were common in public places in the Greek city states. And we know from Geminos (Evans and Berggren 2006) that such mathematicians as Eudoxus (390–338) composed almanacs.

Parapēgmata were based on the motion of the sun, whose annual motion around the zodiac we referred to above. As for length of the year, the mid-fifth-century Athenian astronomer Meton, who also composed a *parapegma*, calculated the length of the solar year as 365 ¼+1/76 days, which was the best value available until almost 300 years later.

Although the year was based on the sun, the months were based on the moon, one month being the time from one full moon to the next. However, the

FIGURE 5.2. Parapegma fragment (456B) from Miletus. Photo © Antikensammlung, Staatliche Museen Zu Berlin-Preussischer Kulturbesitz. Photo: Johannes Laurentius; Inventory number: SK 1606 IV Milet Inv. 456B. This fragment found in the ancient site of Miletus contains parts of three columns. The first begins "The sun is in Sagittarius" and the second begins "The sun is in Aquarius."

length varied between 29 and 30 days, with 29 ½ as a near average. So there was no simple relationship between the lengths of the month and the year until Meton calculated that 19 solar years contained 235 lunar months, seven more months than they should have if all years are to have 12 months. So, Meton inserted one "leap" month in each of seven years over a nineteen-year period. This brought the number of days in nineteen years to within a quarter of a day of the number of days counted according to the numbers in each of the 235 months. After nineteen such years, a new cycle could start all over again. So, if one were not bothered by ¼ day in nineteen years, one could say that the calendar problem had been solved. And about a century later, the astronomer Callippus of Cyzicus (370—300) composed a larger cycle, known as a Callippic cycle, of 76 (=4x19) years to deal with that extra ¼ day. Four hundred years later Ptolemy still used this cycle.

ASTRONOMY: THE FIRST GREEK GEOMETRIC MODELS AND THEIR FRAMES OF REFERENCE

But while numerical modeling was underway, Greek philosophers living in the southern part of Italy and in Sicily introduced a new idea. It was here that the notion of a finite, spherical cosmos centered on a spherical earth originated. Some ancient traditions trace the idea back to Pythagoras of Crotona (*c.* 530), but in any case, it is found in a didactic poem by one of his followers, Parmenides of Elea (*c.* 450). The Pythagoreans revered the sphere as the most beautiful shape and the Greek word *cosmos* comes from the Greek word for ornament (as in "cosmetic").

Although the astronomers believed the cosmos was finite, its size was believed to be so great that the whole earth could be taken as a mere point at its center. This new idea opened the way to geometrical explanations of the phases of the moon, lunar and solar eclipses and, eventually, to the motions of the five naked-eye planets. The model based on this arrangement of a spherical cosmos surrounding a spherical earth has become known as the "two-sphere model." And this model led the study of astronomy in a new direction. It also appealed greatly to Parmenides' friend Plato (*c.* 429–347), who challenged astronomers to account for the motions of the heavenly bodies by means of the rotating spheres and circles.

Apart from philosophical satisfaction the two-sphere model provided three frames of reference for observations and calculations, frames which are still in use today. These are: (1) the equatorial system, based on the north–south polar axis with the equator being the great circle of all points on the sphere equidistant from the two poles; (2) the horizon system, whose reference circle is the observer's horizon and whose poles are the zenith (overhead point on the sphere of the cosmos) and nadir (the point on the sphere opposite the zenith); (3) the zodiac system, whose great circle is that traced out through the heavens by the sun during its yearly journey around the cosmos. All three great circles pass through the observer at the center of the cosmos and are inclined to the celestial equator—the zodiac at an angle of about 23 ½ degrees, but the horizon at an angle that depends on the observer's latitude.

These three systems described above were combined in an instrument modeling the celestial sphere and known as an "armillary sphere," which Ptolemy described in *Almagest* V,1 (Toomer 1984).

In this instrument, the three reference circles were displayed as rings concentric with the central earth on a central axis representing the polar axis (Figure 5.3). Ptolemy used all three of these reference frames throughout his *Almagest* (Toomer 1984), together with a theorem in spherical trigonometry we know as Menelaus' Theorem. Menelaus (*c.* 100 CE), who worked in both Alexandria and Rome, wrote the first extant work on spherical triangles, i.e. triangles whose sides are arcs of great circles on the sphere.

Prior to the discovery of Menelaus' theorem astronomers could measure arcs on the celestial sphere by a technique known as an analemma, now regarded as part of descriptive geometry. In the first century BCE, the Roman architect

FIGURE 5.3. Armillary Sphere. Photo courtesy of J. Lennart Berggren. The three rings represent the horizon (green), ecliptic (red), and equator (blue), and their subdivisions. A circle (purple) parallel to the equator represents the Tropic of Cancer.

Vitruvius describes how to use this technique to construct sundials (Granger 1983: 249–57).

REFINING THE TWO-SPHERE MODEL: THE RISING TIMES AND THE HOURS

In the second century an Alexandrian mathematician named Hypsikles solved an important problem, using arithmetic rather than geometry. The problem was to determine the length of daylight on a given day at a given locality whose maximum length of daylight was known. In Hypsikles' locality this maximum is about14h. In the course of a day the entire zodiac rises and sets, so the problem was to find the sum of the rising times of the six signs that follow the sun as it crosses the sky during the day. Hypsikles made the simplifying approximation that the rising times increase by a constant amount from a minimum value for the signs of Pisces and Aries, which rise most nearly vertical to the horizon, to a maximum value for Cancer and Sagittarius. He then used his mastery of the sums of arithmetic progressions and symmetries of the zodiac to solve the problem. And he went even further to calculate the rising times of each *degree* of the 360° in the zodiac—a virtuoso performance designed to impress other mathematicians.

The Greeks also invented devices to show the hours of the day. Among these devices were water clocks, candles, and sundials. As to sundials, many ancient societies measured the passage of time by the length of the shadow cast by a vertical pole (*gnōmōn* in Greek). And in his *History*, Herodotus tells us that it was from Babylon that the Greeks learned of "the twelve hours of the day and the gnomon and the pole."

Greek astronomers designed the first sundials based on mathematical principles and the earliest known Greek sundial was based on a horizontal rod that cast a shadow on a spherical cavity whose tip modeled the movement of the sun in the sky above it (Figure 5.4). The tip of the rod's shadow indicated the hours by means of curves incised in the concave surface, as though the sky had been projected onto the spherical surface through the tip of the gnomon.

One method of telling time without the sun requires the use of a water-driven clock that Vitruvius called *anaphoric* (Granger 1983: 61). It consisted of a rotating disk bearing a mathematically accurate map of the heavens showing the zodiac and prominent constellations (a fragment can be seen in Figure 5.5). An openwork circular grid in front of that disk displayed an image of the local horizon and curves, allowing the user to tell the hours of the day or night.

Representing the celestial sphere faithfully on a plane surface demanded a mastery of geometry including the use of stereographic projection. This projection was also the theoretical basis of a scientific instrument known as the

FIGURE 5.4. Greek Sundial. Photo by Markus Matzel / ullstein bild via Getty Images.

FIGURE 5.5. Anaphoric clock. Photo courtesy of Salzberg Museum. A line drawing of inscriptions on a fragment of an anaphoric clock (showing four zodiacal signs and a few constellations) from an ancient Roman site near Salzburg. The images of the signs broke along the arc where the holes representing the path of the sun in the zodiac were drilled.

astrolabe, not to be confused with the Babylonian "astrolabe" tablets discussed above. We have no sure evidence of the Greek astrolabe before Ptolemy explained the projection in his *Planispherium* (Sidoli and Berggren 2007), where he explained how to use the projection to make a plane star map of the spherical heavens.

SOLAR ANOMALY, THE MOON AND THE FIVE PLANETS

Callippus, whose 76-year cycle we mentioned above, discovered something the Babylonians had also realized: the four seasons had different lengths. Autumn was the shortest with 89 days and spring was the longest with 94 days. This discovery had serious implications for the two-sphere model because it implied (retaining the hypothesis of circular motion) that either the earth was not at the center of the sun's circular orbit, despite what had been generally believed, or the sun moved around the zodiac at varying rates on different parts of its orbit. For, if its motion were at a constant rate on a circle centered on the earth, the four seasons would necessarily have the same length.

In the mid-second century, Hipparchus resolved this problem with a kinematic model (see Diagram 5.3a), in which the sun moved around its circular orbit at constant speed, but the center of its orbit was displaced from that of the earth by a distance that he calculated to be about 1/24th the radius of its orbit *Alm.* III, 4 (Toomer 1984). Moreover, he calculated that the displacement was in the direction of 5 ½° of the zodiacal sign, Pisces. Hence, the sun would be furthest from the earth at that point of its orbit.

To make these calculations, Hipparchus constructed a table of chords in a circle in which he calculated the lengths of chords subtending arcs of 7 ½°, 15°, . . ., 180° in a circle of radius 60. Some 300 years later, Ptolemy in *Almagest* I, 11 (Toomer 1984) used successive approximations to refine this table to one in which the steps were ½°. Tables of chords are equivalent to our sine table and Hipparchus may be credited with the invention of trigonometry. As another contribution to the motion of the sun, Hipparchus shortened the solar year to 365 1/4 - 1/300 days, a figure that Ptolemy—after much discussion—took as the basis for his solar theory in Book III of his *Almagest* (Toomer 1984: 139–40).

Ptolemy adopted Hipparchus' kinematic model for solar motion, and in constructing his tables of solar motion Ptolemy said he would use the mean motions he had derived to compute solar position back to the time of the Babylonian king Nabonasser (787), since "that is the era beginning from which ancient observations are, on the whole, preserved down to our own time" (Toomer 1984: 166)

Hipparchus' solar theory brings out a major difference between Babylonian and Greek astronomy. Whereas the Babylonians wanted to know when an

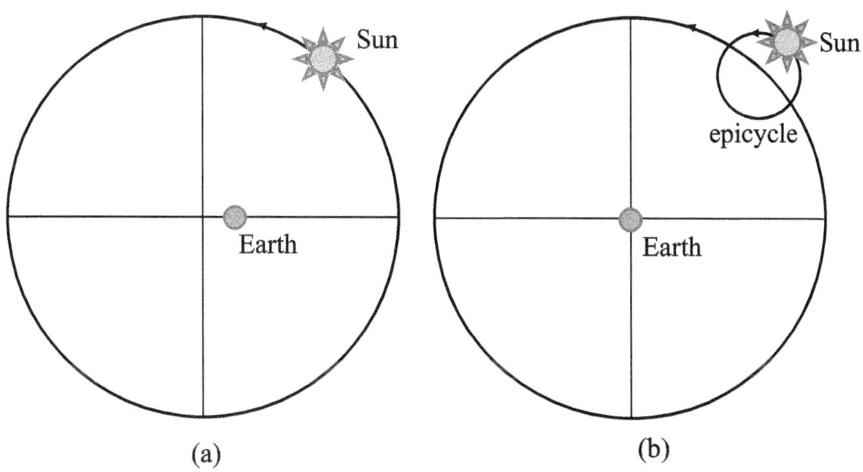

DIAGRAM 5.3. Hipparchus' two models of solar motion. In both, we are looking down from the north at the sun's orbit in the plane of the ecliptic. Earth is off-center in model (a) and the sun moves counter-clockwise around that center. In model (b) the sun travels around an epicycle whose center moves around the large circle, its deferent, centered on the earth. All rotations are uniform.

event, e.g. a solar eclipse, would occur, the Greeks wanted kinematic models that would answer where in the zodiac the sun, moon, or a planet would be on any given day. Less than a century earlier, Archimedes had given a physical version of such modeling, where one could see the celestial objects in continuous motion. Hipparchus' theory made it possible to construct numerical tables for the day-by-day motion of the sun throughout its annual cycle.

His model of lunar motion was constructed from data on three lunar eclipses but was not a general theory. For that he needed to deal with three problems. First, as Aristarchus had shown (Heath 1913), unlike the sun, the moon was near enough to the earth that the difference in its position among the stars as seen from two localities, say Rome and Alexandria, was on the order of 1/10°. (This gives some idea of the kind of precision astronomers sought in the second century BCE.) Second, as the moon travels around the zodiac it does not stay on the ecliptic as the sun does. Rather, it snakes around by about 5° on either side of the ecliptic, a motion the Greeks compared to that of a dragon. Third, the motion of the moon was irregular from one day to the next, sometimes swifter but on other days less swift. The motion was, as the Greeks put it, anomalous.

The importance of solving these problems was that knowing the position of the moon one can then find the positions of stars. Hipparchus made the first catalog of the stars and Ptolemy revised and extended that work in *Alm.* VII (Toomer 1984). In looking at older reports of the positions of the stars, Hipparchus found some discrepancies between his observations and older reports, sometimes amounting to as much as 2°. Differences of this magnitude could not be ignored and Hipparchus concluded that the N–S axis of the cosmos rotates from west to east around the axis of the ecliptic.

Consequently, the points where the equator intersects the ecliptic (the two equinoxes), must move very slowly eastward along the ecliptic at a rate Hipparchus reckoned to be about 1° per century (see Diagram 5.4). Although this may not seem like much, it does add up. This phenomenon is known as the precession of the equinoxes, and the actual rate is about 1° every 72 years, so that in a little less than 26,000 years, the equinoxes and solstices travel completely around the zodiac. The discovery of precession has allowed astronomers to relate ancient observations to modern positions, and Hipparchus' work in general brought much greater precision to astronomical calculations and a powerful mathematical tool, trigonometry, to use in those calculations.

The mathematical description of the motions of the five planets is the last refinement in Greek astronomy we shall discuss here. The planets Mercury, Venus, Mars, Jupiter, and Saturn were the only ones visible to the Greeks, and their erratic motions gave them their Greek name, *planētes* (= wanderers). The chief difficulty with modeling the motions of the planets was that in addition to

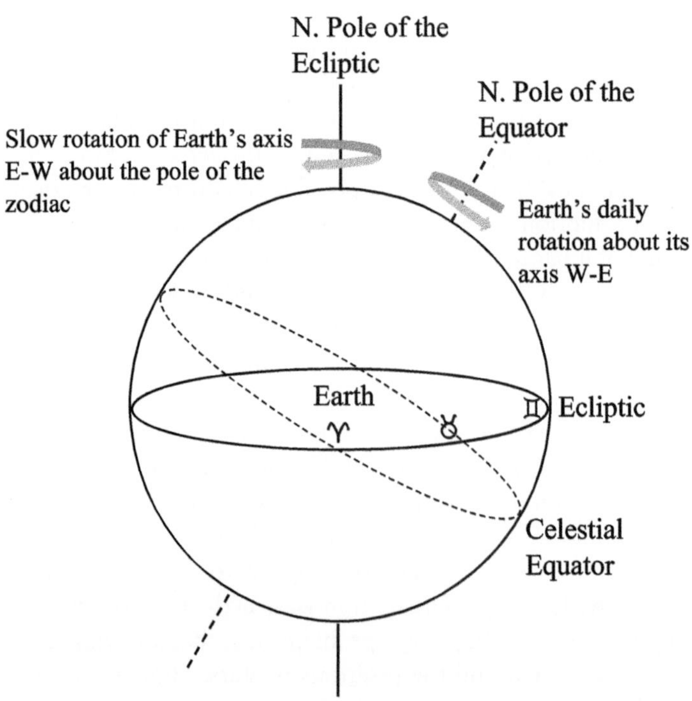

DIAGRAM 5.4. Precession. As the earth's N–S axis rotates slowly around the N-pole of the ecliptic the intersection of the equator with the zodiac moves eastward through Aries, Taurus, Pisces, etc.

sharing the moon's problems of not moving at a constant rates and not staying on the ecliptic, they did not even move always in the same direction. Plato likened them to dancers—always going back and forth!—and this backward motion (called "retrogade") challenged astronomers to find a model that would account for their erratic motions but would not violate the requirement of uniform circular motion.

By numerical models the Babylonians were able to predict with considerable accuracy the times and places in the zodiac when any one of the five planets would halt and reverse its motions and the length of the cycle over which the pattern of back-and-forth motions for any given planet would repeat itself. But this is not the same as answering the question, "Where will Mars be on any given day?"

The Greeks took over these period relations from the Babylonians, and Plato's friend Eudoxus of Cnidus tried to model the motions of the planets by using a geometrical model of nested spheres. (Although our diagrams will show the planets moving on circles, the Greeks thought of them as attached to rotating spheres and spoke not of the "orbit of Aphrodite" (say), but of "the sphere of Aphrodite.")

The model required four nested spheres for each planet, each of the three inner spheres rotating around an axis joining it to its neighboring outer sphere (see Diagram 5.5). And each of the four spheres rotated around its own axis at a different rate. For certain planets, these models produced retrograde motion by means of the two innermost spheres (solid). Their combined rotations made the planet move on a curve the Greeks called "hippopede" (horse-fetter), an elongated figure-8 turned on its side, but the resulting motion did not agree with observations. Nevertheless, Eudoxus' models showed that retrograde motion could be produced by uniform circular motions. (For a detailed account of Eudoxus' models, see Evans 1998: 305–12.)

A more fruitful lead came with Hipparchus' observation that, in modeling the solar motion, it makes no difference (see Diagram 5.3) whether one assumes the sun moves around the earth in an eccentric orbit or that the sun moves around a small circle whose center, in turn moves around a circle centered on the earth. Such a small circle was called an "epicycle" and two approaches, eccentric or epicyclic, produced the same orbit for the sun.

But when Ptolemy applied Hipparchus' epicyclic model to the planets in *Alm.* IX–XII (Toomer 1984), he made two refinements. For each planet, P, he moved the observer, O, a certain distance away from the deferent center, D, and introduced a new point, E, on OD extended beyond D so that ED=DO (see Diagram 5.6). He called E the equant point and the center of the epicycle moves along the deferent uniformly as seen from the equant, not with respect to the deferent center. It is angle SEC, not SDC, that grows uniformly.

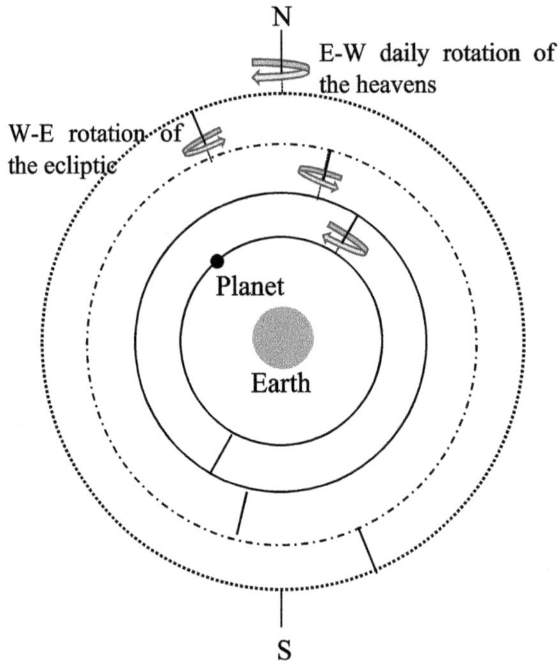

DIAGRAM 5.5. Eudoxus' model for the motion of a planet. This figure shows a section of the celestial sphere by a plane passing through the center of the earth and the two poles.

MEASURING THE COSMOS IN THE CLASSICAL WORLD

The two-sphere model provided not only explanations of why we observe the phenomena we do but, combined with developments in Greek mathematics, it gave a means of calculating the sizes and distances of the sun and moon and even estimating the size of the cosmos. Aristarchus (*c.* 310–230) discusses these topics in his treatise *On Sizes and Distances* (Heath 1913) and Archimedes (287–212) in his *Sand Reckoner* (Heath [1897] 2019).

Aristarchus' treatise uses evidence gathered from lunar observations and the logical model of exposition in Euclid's *Phenomena* (Berggren and Thomas 2001) and *Elements* (Heath 1956), i.e. initial hypotheses, a statement of what was to be proved, and a deductive argument based on the hypotheses to prove it. Aristarchus' arguments were geometrically sound but the data he used were badly off. In particular, one example being his Hypothesis 6: the apparent diameter of the moon is 1/15 of a zodiacal sign, i.e. 2° (rather than the correct value of ½°). This made the distances much smaller than they should have been. But, despite these shortcomings, Aristarchus gave the first demonstration of

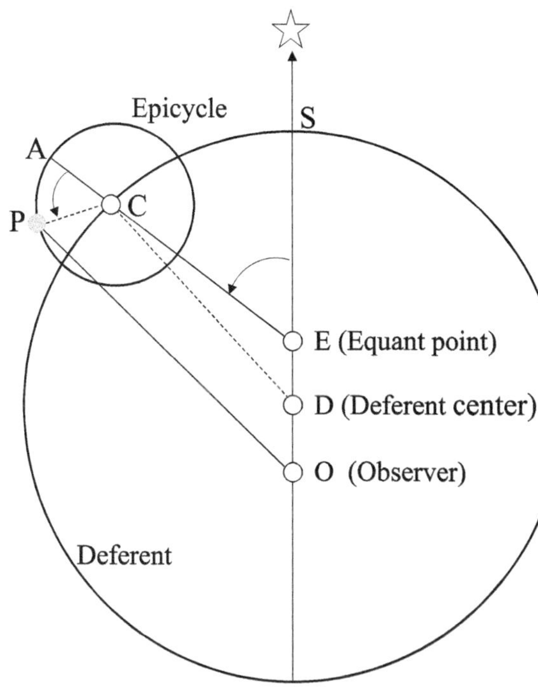

DIAGRAM 5.6. The equant. The line ODES points to a fixed star and C, the epicycle center, moves uniformly around E with respect to that line. And P moves uniformly around C with respect to line ECA.

how one could use Greek geometry and astronomical data to attack difficult problems.

Archimedes' *Sand Reckoner*, was devoted to showing that the number of grains of sand in a sphere the size of the cosmos was not only finite but could be expressed in words. As was mentioned in Chapter 1, Greek number notation did not lend itself well to expressing very large numbers, so Archimedes invented a system to express conveniently numbers on the order of 10^{63}. Thus, supposing the cosmos was a sphere, it was necessarily finite and so one could ask how big it was. Dedicated to a reigning monarch, the treatise was clearly intended as a mathematical showpiece.

Archimedes, however, not only provides a new way of expressing very large numbers, but also gives us the first description of an instrument especially constructed to make an astronomical observation—in this case, the apparent diameter of the sun. And, as did Aristarchus, he shows how one can use geometry to derive conclusions from astronomical data. His observation showed that (expressed in modern terms) the visual angle subtended by the sun was less than 0.55° but more than 0.45°. So, taking the average, he arrived at ½°, which, he says, Aristarchus had already determined.

GREEK MATHEMATICAL GEOGRAPHY

The same two-sphere model that shaped the course of Greek astronomy also shaped the approach of Greek mathematical geography. By this we mean the study of geography as it used mathematics and astronomy to describe the positions of localities and to construct plane maps faithfully representing the lands and seas of the known earth. Indeed, Ptolemy devoted his *Geography* (Berggren and Jones 2000) to exactly these tasks. (Henceforth, "map' will refer to a plane map of a sphere.) In fact, Hipparchus had stressed that astronomy provided a framework for scientifically locating points on the earth's surface.

We have already described the idea of equatorial coordinates in astronomy and it was the projection of the celestial equator and its parallels onto the surface of the earth as a reference frame that brought this system down to earth. The parallels then gave rise to geographical latitudes and the meridians to longitudes.

A century earlier, Eratosthenes (*c.* 275–194) of Alexandria, a contemporary of Archimedes, had calculated the size of the earth using mathematics and three data: (1) the Egyptian cities of Alexandria and Soēnē (modern Aswan) lay on the same meridian; (2) on the summer solstice, the noon sun is directly overhead in Soēnē but about 1/50 of a circle south of the zenith in Alexandria; and (3) the distance between Alexandria and Soēnē is 5,000 *stade*s (see Diagram 5.7 for Eratosthenes's procedure). So, by parallel lines, since the angle made by the sun and the vertical in Alexandria is equal to the angle between Alexandria and Soene at the center of the earth, he multiplied 5,000 *stade*s by 50 to obtain the circumference of the earth as 250,000 *stade*s. This he rounded to 252,000 *stade*s so as to arrive at the figure divisible by 60.

Hipparchus accepted Eratosthenes' figure for the circumference of the earth and its consequence, that 1° of the earth's surface N–S along a meridian or E–W along the equator was equal to 700 *stade*s. And he calculated certain celestial phenomena for a number of localities along the meridian of Meroē, a royal city on the Middle Nile. Among such phenomena are: (1) the height of the pole star above the horizon (which is equal to the latitude); (2) when (if ever) the sun is in the zenith; (3) the ratio of lengths of equinoctial and solsticial shadows to the length of a vertical rod; and (4) the length of the longest day. And in *Alm.* II, 6 (Toomer 1984), Ptolemy tabulates, for every ¼ hour increment of the length of the longest day, beginning with 12 hours at the equator, both the celestial

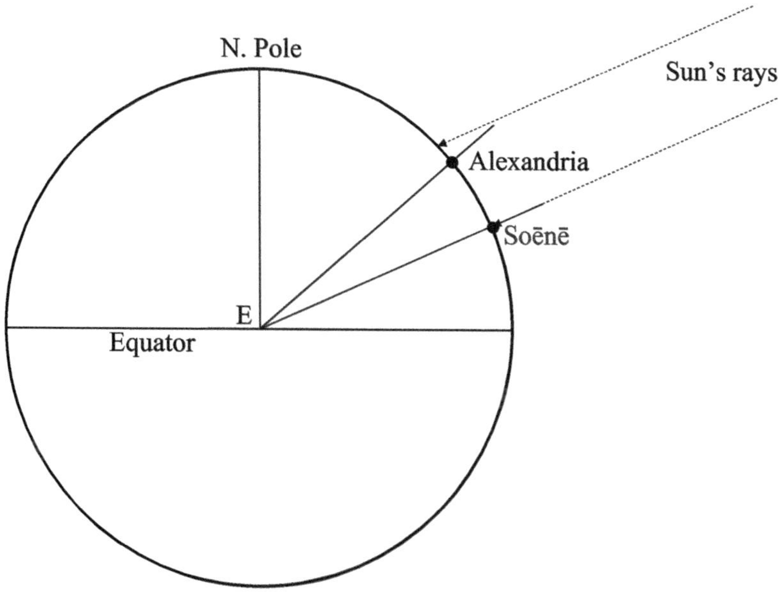

DIAGRAM 5.7. Eratosthenes method for measuring the circumference of the earth. The circle is the common meridian Eratosthenes assumed for Alexandria and Soēnē. Observations and geometry showed that the arc joining the two cities was 1/50th of a circle—and since the sun's rays were assumed parallel this was the same angle made by the sun's rays and gnomon at Alexandria. Records gave the distance between the two cities as 5,000 *stades*, from which the circumference of the earth could then be found.

phenomena mentioned above and known localities or geographical features on the various parallels.

Eratosthenes made a map of the known world but the first cartographer to make a map based on parallels and meridians was Marinus of Tyre, who lived perhaps fifty years before Ptolemy. His map was rectangular, and meridians and parallels were represented by mutually perpendicular straight lines (Diagram 5.8).

Such a grid was not hard to construct but it represents the distance between two meridians as the same for all latitudes, which is not what one sees on a globe. This violated what Ptolemy refers to in his *Geography* as the "proportionality" a map should have and he criticizes it sharply as a scheme "the least likely to preserve proportionality."

One thing Marinus did to keep some proportionality was to scale the spacing between meridians of the grid so that a degree on the parallel of Rhodes was 4/5 the length of a degree on any meridian. Since this is very close to reality, and many localities were not too far away from this latitude, distances and directions for many important places on the map modeled those on the earth reasonably

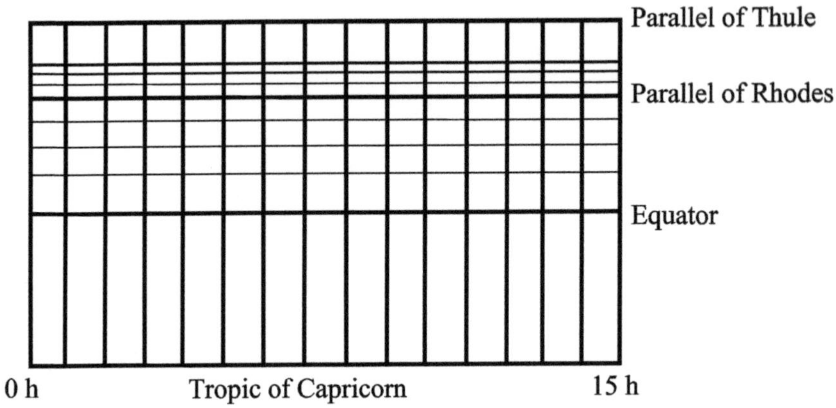

DIAGRAM 5.8. The grid for Marinus of Tyre's map. The grid represents the latitudes as parallel horizontal lines and longitudes as parallel vertical lines. The unequal spacing of the latitudes is due to the fact that the latitudes chosen are boundaries of broad geographical regions known as the *klimata* (from which we derive the word "climate").

well. And, although Ptolemy levels a number of criticisms on several features of Marinus' map, it is also clear that he made heavy use of Marinus' work for his own map.

Among the uses were: deciding which localities lay on a given parallel or meridian; locating localities from N–S according to *klimata* (see above); retaining the 4/5 scaling for the parallel of Rhodes. And he imposed similar requirements for lengths of the parallels through Thule (the northernmost boundary of his map at 63° N) and the equator.

When Ptolemy wrote his *Geography* Roman armies, merchants and travelers had explored the world from the Ganges to the Straits of Gibraltar and from areas south of the equator to the lands of the far north. About this vast expanse, Ptolemy felt he had enough data to make a "proportionate" map.

Ptolemy clearly felt that by organizing cartography around latitude and longitude he would achieve this goal, and he published two maps to do so. In his first projection (Diagram 5.9), his goal was to make a map that, as the eye passes over successive meridians, it sees something similar to what it would see if it were to survey a globe from E–W. So, the meridians are straight lines fanning out from a common center, just as meridians on a globe fan out from the North Pole. And, as on a globe, that point is also the center of circular arcs representing latitudes on the map.

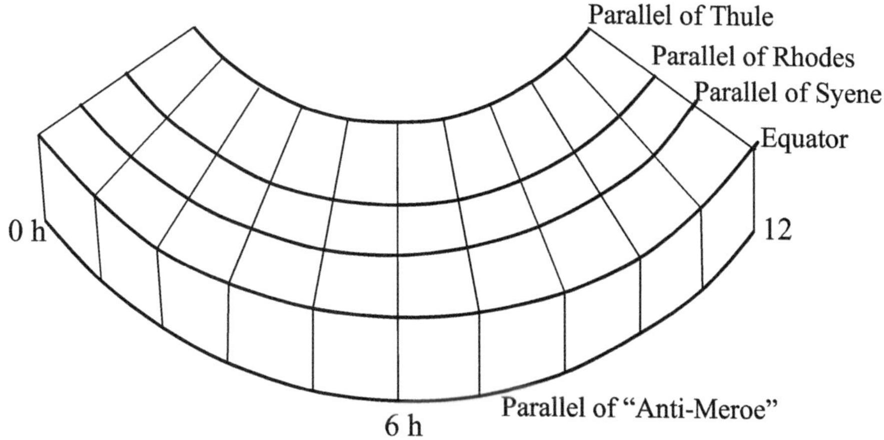

DIAGRAM 5.9. Ptolemy's first projection. Ptolemy reduced the longitudinal dimension of the *oikoumene* (i.e., the known inhabited world) by 20% of Marinus' dimension of 15h. He also moved the southern boundary about 8° north of Marinus' boundary of the Winter Tropic.

Some have criticized Ptolemy for the abrupt jog inwards that the meridians make at the equator. But, in terms of his goal in the first projection of preserving visual appearance as simply as possible, a sharp turn is understandable.

In Ptolemy's second projection, he does as much as he can to imitate what the eye would see, looking towards the center of the globe from above the central point of the *oikoumene*, which he takes to be the intersection of the parallel of Soēnē (the modern Aswan) with the meridian of the southeast tip of the Arabian Peninsula (Figure 5.6).

Viewing the map from above this central point one would see parallels of latitude, and all meridians but the central vertical line, as arcs, for which Ptolemy chose circular arcs. Although he surely knew what Greek geometers had known for centuries—that these arcs would have been seen as ellipses—it seems he had no desire to add to the cartographers' labors. This second projection set the "gold standard" for world cartography, and it remained so for over 1,400 years.

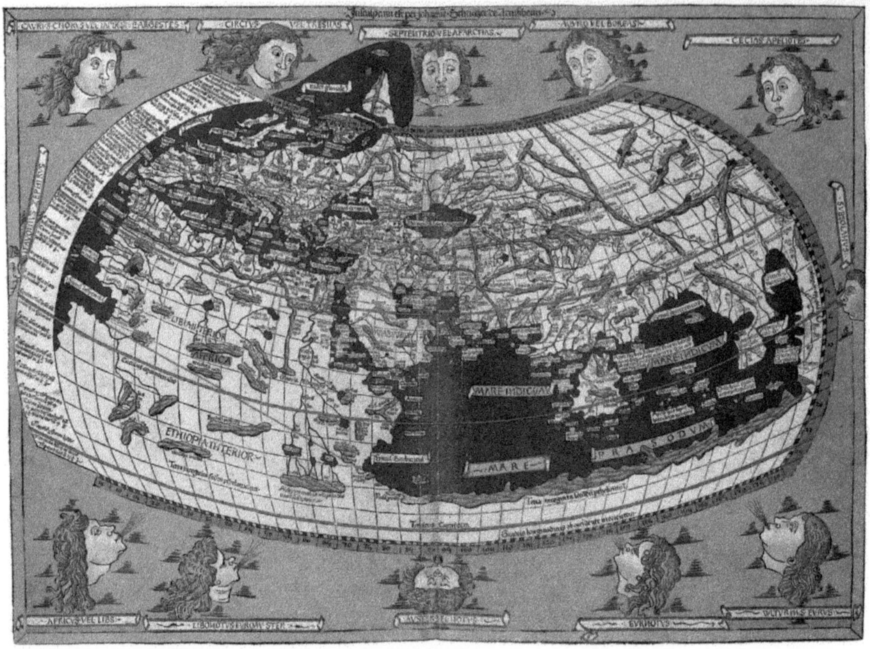

FIGURE 5.6. Ptolemy's Second Projection from his Geography. Photo by Wikimedia Commons. Ptolemy's second projection is shown in the splendid edition of his *Geography* published in Ulm in 1482. The figures of the winds blowing, with the name of each wind next to its head, refers to a common ancient and medieval method of indicating directions.

CHAPTER SIX

Mathematics and Technological Change

Technical Procedures and Numerical Knowledge in Cuneiform Cultures

EDUARDO A. ESCOBAR

The cultural relationship between mathematics, numeracy, and technological knowledge in the ancient world is bound directly to how we define "technology" itself. Today, "technology" can be counted among the most plastic terms in the English lexicon, and can refer to anything from machinery, language, and communication, to religion. In modern usage and within scholarly discourse, "technology" and its history refers generally to the industrial arts, craft techniques for manufacturing, and/or the applied sciences. While defining technology in terms of applied science (or applied mathematics) may seem intuitive, it is also historically recent, exemplified by the twentieth century development of the atomic bomb, which in turn led to a conceptual shift that reinforced the link between "technology" and "applied science" (Schatzberg 2018: 12–13).

Conceptualizing technology as dependent on, or as the "handmaiden" of the sciences (physics and mathematics in particular), well predates the Second World War and even the term technology itself. We can trace the Classical world, wherein we encounter clear boundaries between craft and knowing-*how*, i.e., *technē*, and the mental process of reasoning and abstraction and knowing-*that*, i.e, *epistēmē*, but it is crucial to emphasize that this dichotomy is neither universal nor even applicable to non-Western cultures. Overcoming the inherited boundary that separates the head from the hand is challenging precisely because it is deeply cultural, and continually reified in contemporary scientific practices that are keen to separate fields into their "practical" and "theoretical" branches. Take, for example, our most advanced scientific instrument to date, the Large Hadron Collider; it is home to an international community of scientists, many of them physicists who fall into one of two clans. On the one hand, there are the *theoretical* physicists and on the other hand, the experimentally minded *applied* physicists.

The boundary between the practical and the theoretical is reified not only in modern physics, but in the writing of history itself. Histories of ancient technology (*technē*) have traditionally focused on the reconstruction of ancient craft practices and somatic know-how principally employing archaeological evidence and secondarily, texts which lend support to the reconstruction of said craft practices. Histories of ancient science, in contrast, are principally textual; they deal in numbers, axioms, and abstract claims (*epistēmē*) made by literate scholarly communities. Resultingly, cultural conceptions of technology and, in particular, how textual communities understood the transmission of technological knowledge—the subject of the present chapter—is a subject that remains underexamined.

What would a brief history of technology look like if it set aside the question "how did the ancients make things?" to focus on the question of "how did ancient textual cultures codify and conceptualize what we call technology?" Using cuneiform technical procedures composed between the second and first millennia BCE, this chapter aims to provide a brief history of technological knowledge transmission as represented by cuneiform cultures of the ancient Middle East. What follows explores how, for cuneiform scribes, the transmission of technological knowledge was an inherently scholarly endeavor, closely bound with the linguistic framework of Old Babylonian mathematical procedures, and historically dependent on knowledge of numeracy and mensuration.

CUNEIFORM TECHNICAL PROCEDURES: MAKING AND CALCULATING

Writing cultures, globally and throughout time, have independently invented the instruction manual. But there is no universal consensus as to how instruction manuals codify, preserve, and transmit knowledge, not even within the same set of cultures. This condition—that of the cultural contingency of instruction manuals—is precisely what makes Akkadian procedural texts so pertinent for examining the cultural history of mathematical and technological text traditions. An Akkadian glassmaking recipe from the Neo-Assyrian period in the seventh century is thematically, socially, and epistemically distinct from a thirteenth-century Middle Assyrian procedure for training horses. And yet, both share the same linguistic register, including syntax, voice, and shared approach towards the generation of technical language and exactitude that was first initiated in Old Babylonian mathematical contexts during the nineteenth century BCE. Indeed, in terms of the *longue durée*, the history of the Akkadian instruction manual begins with cuneiform scholars who first sought to encode and transmit mathematical knowledge.

Cuneiform procedures—composed principally in Akkadian (a semitic language)—are instructional texts that provide "you," the grammatical subject, with step-by-step instructions for a variety of technical subjects. Cuneiform procedures and recipes treated topics such as mathematical problem-solving, glassmaking, horse-training, perfumery, medicine, ritual purification, and astronomy. The geographic expanse of cuneiform procedures follows the spread of the script itself, and includes the regions of modern-day Iraq, Syria, Turkey, and Iran. Technical procedures for the training of horses are particularly productive for considering the geographic and historical transmission of technology and numeracy; these same texts, written using the cuneiform script by both Hittite and Akkadian scribes, often employed foreign technical terms. In addition to their thematic, geographic, and linguistic breadth, procedural texts are also among the most historically long-standing categories of scholarly texts, covering a period of two millennia, with the earliest exemplars dating to the early second millennium BCE. Among the procedures of special interest to the topic at hand—namely, those concerning the scholarly intellectual history of mathematical and technological knowledge—are Akkadian recipes for making colored glass (*c.* seventh century BCE), Hittite and Akkadian procedures training horses (*c.* thirteenth century BCE), and the so-called Esagila tablet (*c.* third century BCE), which provides instructions for calculating the historical dimensions of a sacred Babylonian structure. Collectively, this broad selection of texts serves to highlight the degree to which cuneiform procedures across millennia demonstrate a sustained interest in codifying ancient number terms, *termini technici*, archaic systems of metrologies, fractional (often miniscule) quantities of ingredients, and timekeeping within ritual craft contexts. These shared thematic components, in addition to the linguistic framework discussed below, aid in unifying the technical procedures under consideration.[1]

A number of Akkadian texts self-identify as "procedures" using terms like *nepešum* and *epištu*, both derived from the verb *epēšu* "to make."[2] Although the verb *epēšu* has the base meaning "to make," it can also mean "to compute" in technical contexts. In Akkadian, making and computing form part of the same semantic family. Variations of the procedural phrase, "you in your making/computing," *atta ina epēšīka*, can be found on Old Babylonian mathematical problem texts from the early eighteenth century, Neo-Assyrian glassmaking recipes from the seventh century, and Late Babylonian astronomical procedures from the second century BCE. Expanding upon these linguistic and historical origins, this chapter explores case studies from cuneiform cultures that attest to the interaction and co-dependence of numerical knowledge and technology, shedding light on the scholarly mechanisms and motivations that perpetuated the production of technical procedures for two millennia of intellectual history. As with any historical text, cuneiform procedures were subject to cultural variations. However, the incredible longevity of the cuneiform procedural text format points to a concerted effort by scribes to codify and transmit technical and numerical knowledge within a set of literate cultures often overlooked by general histories of technology. In this chapter, the following terms are employed with frequency: "procedures," understood as a broad category of instructional texts written using second-person verbal forms, and which include technical "word problems" in mathematical contexts; and "recipes" such as glass- and perfume-making texts in late second and mid-first-millennium contexts. I employ the phrase "cuneiform cultures," following Radner and Robson, to emphasize "cuneiform literacy and the literate segments of society, or 'textual communities'" across the ancient Middle East (see Radner and Robson 2011). It is precisely the linked intellectual histories of technology and mathematical knowledge within the *longue durée* of "cuneiform cultures" that is at stake in this work.[3]

As a result of the durability of the preferred writing medium for cuneiform scribes—the clay tablet, a technology in its own right—Assyriologists are in a unique position to trace the rich intellectual history of technological and mathematical knowledge in the ancient world. With a rich set of textual sources in hand, the goals of this chapter are twofold. First, the majority of the chapter will present an episodic history of cuneiform technical procedures, focusing on language, epistemology, and the socio-intellectual contexts of texts under discussion. Herein, we will focus on how cuneiform scribes packaged technological and mathematical knowledge within a procedural framework and consider why the textual format proved to be so successful across two millennia. Second, the chapter will conclude with an evaluation of procedural knowledge and return to the question that initiated this inquiry, namely: what can cuneiform procedural texts offer in terms of expanding the conceptual scope of technological and mathematical knowledge transmission?

PROCEDURAL LANGUAGE AND TECHNICAL KNOWLEDGE IN CUNEIFORM CULTURES

Language represents the clearest starting point for this episodic survey of procedural knowledge. Among the most historically long-standing phrases found within procedural tradition is the Akkadian expression (*atta*) *ina*/*ana epēšika* "(You), in your making/calculating/computing." As expected over the course of two millennia, the phrase underwent a number of orthographic changes and witnessed a range of local variations. Thematically, however, our current evidence indicates that this particular procedural phrase was limited to texts concerning mathematics, glassmaking, and astronomy.

As scholars like Eleanor Robson have noted, that the earliest and most well-attested iterations of the phrase originate in the northern Babylonian Diyala region, also known as the Eshnunna region (Robson 1999: 8). After the late second millennium BCE, the phrase is abbreviated and lost the initial pronoun "you" (*atta*), which characterized the Old Babylonian problem texts from Eshnunna and surrounding region.[4] Neo-Assyrian glassmaking recipes "for your making lapis lazuli" preserve among the few early mid-first-millennium attestations of the phrase. Centuries later, following the collapse of the last native empires in Mesopotamia, *ina/ana epēšika* is once again utilized in the context of Late Babylonian astronomy. Compellingly, within these linguistic contexts, we find the double-semantic sense of *epēšu* "to make/calculate" clearly expressed in an enigmatic set of procedures for making a gnomon (sun dial), which begin: "for you to make a 'crescent.'" Here, the semantic ambiguity of *epēšu*, understood as both making and doing, is especially effective, as it reflects both the physical, fabricated properties of the instrument, as well as its function as a calculating device.

A variation of the phrase, using the same syntactic structure as *ana epēšika*, is the phrase *ana amarīka*, "in your observing." Preceded by the conditional particle *šumma*, "if," the phrase is also translated, "if you are to observe," as in the following example from the astronomical compendium MUL.APIN:

> *šumma mūṣē šārī* ***ana amārika*** *Ereqqu ina tīb iltāni parkat*
>
> If ***you are to observe*** the direction of the winds: the Wagon flies across where the North wind rises . . .[5]

The verb *amāru*, "to see," much like the previously discussed *epēšu*, "to make, compute," is a common Akkadian verb with a broad semantic scope. In the linguistic context of a technical procedure, "seeing" takes on a range of meanings. The phrase *ana amarīka* may mean, "for you to calculate/discover/observe/work out." As Ossendrijver has noted, in astronomical procedures from the Late Babylonian period, "[t]here does not appear to be a difference between 'constructing' (*epēšu*) and 'seeing' (*amāru*) a quantity" (Ossendrijver 2012: 36). As we will find within the procedural text tradition (e.g., in the discussion of the Esagila tablet below), calculating, computing, seeing, making, and working, were semantically linked and conceptually codependent.

The earliest period of technical procedures can be set in the south of modern-day Iraq in the region known as Babylonia *c*. 1800 BCE, i.e., the "Old Babylonian" period. While the survey begins in the early second millennium BCE, special attention will be given to technical procedures that were produced between the late second and first millennium BCE that provide exemplary evidence for the employment of numeracy and procedural knowledge in technological contexts. These episodes include late-second-millennium Assyrian and Hittite horse-training, seventh-century Assyrian glassmaking recipes, and "The Esagila tablet," a third-century BCE text compiling procedures containing historical dimensions of sacred architectural structures. As this is not a comprehensive study of cuneiform procedural knowledge, the substantial corpus of ritual and medical procedures, which do not consistently employ numerical or metrological data, will not be discussed in any depth.[6] Nor will this chapter cover the Late Babylonian astronomical procedural texts, as these are discussed in Chapter 5 of this volume. Moreover, literary texts which included procedural and even numerical information—as in the Old Babylonian flood story—will also be set aside in order to focus fully on the particular case studies of technological procedures which have seldom been compared within a single study.

THE LEGACY OF OLD BABYLONIAN MATHEMATICAL PROCEDURES

Old Babylonian (*c.* 1800 BCE, henceforth, OB) problem texts begin with scenarios that we might broadly characterize as "practical," and, by extension technological. OB mathematical problem texts can include metrological conversions and land inheritance calculations to technical challenges like digging a trench, calculating the number of bricks it will take to construct a wall for a house, or working out the daily labor rates. Particularly during the OB period, mathematical procedures also preserve a dialectical framework, wherein procedure is initiated by an arithmetic question in the form: "if someone asks you, saying . . ." Once the question is posed, the grammatical subject changes to a generic "you," the anonymous Babylonian student; in IM 053961, a text from Tell Harmal (in the Eshnunna region), for example:

> If someone asks you, saying this:
>
> "A mud wall. The breadth is two cubits, the height one cubit. 'What is the daily work-rate of one man?'"
>
> ***You, in your working***: multiply 2 by <0;05>, your height, so that 0;10 comes up. The 0;10 that come up for you: solve the reciprocal of 0;10 so that 6 comes up. Multiply the 6 that came up for you by 0;03 45, your constant, so that 0;22 30 comes up. The 0;22 30 that came up for you is the daily work-rate of one man.[7]

Notions of working and computing, i.e., derivations from the verb ***epēšu*** are preserved throughout the OB math procedures. Above, the phrase "you, in your working" (*atta ina epēšīka*)—to which we will return later in the context of glassmaking—is common to mathematical procedures from Eshnunna. In contrast, farther south, in the city of Sippar, OB mathematical procedures were self-identifying as "procedures," in Akkadian ***nēpešum***, while still retaining the second-person subject:

> A wall of baked bricks. The width is 2 cubits, the height 1 rod, the baked bricks 9 *sar*. What is the length of my wall? You: solve the reciprocal of 2;15, the constant of a wall. You will see 0;26 40. Multiply 0;26 40 by 9 *sar*, the baked bricks. You will see 4 (the volume). Keep (it). Solve the reciprocal of 12, 1 rod, the height of the wall. You will see 0;05. Multiply 0;05 by the 4 that you are keeping. You will see 0;20 (the position). Solve the reciprocal of 0;10, the width; multiply by the 0;20 that you saw. You will see 2. The length is 2 rods. **The procedure (*nēpešum*).**[8]

Regional variations were ubiquitous during this early phase of Akkadian procedures, and mathematics played a formative role in the linguistic developing the text category. In cities like Eshnunna and Sippar, scribes could take a practical subject matter—e.g., finding the length of a wall or computing work rates—and assign to it a reciprocal computation, a task typically reserved for a tabular text format (reciprocal tables were among the most common school exercises). A number of OB procedures were also composed using imperative verbs, not unlike modern cookbooks, but the imperative would eventually be replaced by second-person present verbal forms, which we find ubiquitously in subsequent periods. Linguistically, the instruction manual was still in its experimental phase during the early second millennium BCE, and it would take another few centuries for Akkadian procedures to employ a standardized format.

While the subject matters for mathematical procedures were inherently technological, insofar as they regard logistical and economic challenges related to urban construction, they are not strictly technological craft instructions. That is, OB mathematical procedures may quantify labor costs, materials, and land mensuration for building a house, but they do not provide instructions for how to build that house. Follow these instructions verbatim and "you" will be left with a pile of bricks and a sizeable team of impatient workmen.

Crucial, however, is the historical legacy of mathematical problem-solving texts, which, following the OB period, set in motion the linguistic and epistemic framework for transmitting and encoding mensuration and decision-making protocols at the heart of both making and computation. The case studies that follow examine how procedural language which originated in the OB mathematical contexts manifested in technological instructions for horse-training, making glass, and computing ancient architecture.

ANCIENT NUMBERS: HORSE-TRAINING PROCEDURES FROM THE LATE SECOND MILLENNIUM BCE

The late second millennium BCE—also known as the Late Bronze Age—witnessed a surge of new technological, social, and political changes.[9] Characterized as an "International Age," during which Akkadian became the diplomatic *lingua franca* in the Middle East and Eastern Mediterranean, the period saw the textual codification of a number of technical subjects, including procedures for perfumery, glassmaking, and horse-training. Unlike the mathematical procedures from the OB period, which were locally produced and archived in private household archives, the instructional texts of this period were held in the archives of major state capitals. When and where technologies like horse-training were first encoded tells us a great deal about the function and status of these technical skills, as well as how "technology" was conceptualized within scholarly frameworks.

Cuneiform technical procedures for training horses, composed in both Akkadian and Hittite, are prevalent at two sites to be discussed below: the Assyrian capital of Assur, and in Southern Anatolia, in the Hittite capital of Hattuša. The late second millennium saw two notable and innovative changes to technical procedures. The first was the standardization of context specific metrologies: a case in point are the horse-training procedures discussed below containing linear distances, or "field lengths," and their subdivisions. The second innovation was lexical and included the introduction of foreign (and ancient) Indo-European numerals terms, which are exclusive to the Hittite corpus of horse-training instructions.

While the horse had been known in Mesopotamia since the end of the third millennium BCE, specialized knowledge of horse-training, including endurance and chariot warfare, was first codified during the Late Second Millennium BCE.[10] Cuneiform horse-training procedures are attested in two scribal traditions. The first comes from the Southern Anatolian Hittite capital of Ḫattuša and includes the famous "Kikkuli text," widely recognized as the earliest horse-training instructions, so-called after the Mitanni trainer by the same name. The Kikkuli text is composed of four tablets and a fifth fragment that provides instructions for acclimatizing and grooming horses over a period of seven months. Importantly, these instructions also preserve rare Indo-Aryan numerical terms for linear distances to be discussed below.[11] The second corpus was discovered at the Assyrian capital of Assur in northern Iraq. Notably, the Assyrian corpus is far more fragmentary, but it, too, preserves numerical and technical instructions for conditioning horses and demonstrates clear thematic and linguistic overlaps with the Hittite textual tradition. Although Hittite and Akkadian belong to different linguistic families (Indo-European and Semitic, respectively), both cultures composed horse-training procedures on clay tablets using the cuneiform script. Moreover, both sets of corpora were ultimately housed in Hittite and Assyrian state capitals, respectively, pointing to the fact that by the late second millennium, technical and numerical expertise had shifted away from the household scribal schools of the earlier OB period, and was now under the purview of state archives. Collectively, both corpora provide unique and compelling evidence for the interaction of language, numeracy, and the textual transmission of technological knowledge.

Although both Hittite and Assyrian texts codified technical knowledge for the training and conditioning of horses, the knowledge presented is neither Assyrian nor Hittite in origin. As is made particularly clear in the Kikkuli text, this equestrian knowledge was passed down from the earlier Mitanni culture, which reached the height of its territorial power in northern Syria between the late fifteenth through fourteenth centuries BCE. Mitanni culture, with its capital situated in the northern Syria, was nestled at the crossroads of two competing powers, the Hittite territorial state to the north, and an ambitious Assyrian state to the east. Therefore, the Hittite horse-training procedure that begins, "Thus speaks Kikkuli, the horse-trainer from the land of Mitanni," is in fact a thirteenth-century Hittite document relating fifteenth-century Mittani knowledge. Equally, the thirteenth-century Assyrian horse-training manuals intimate a close link to this earlier Mittani equestrian tradition, as will become clear when both procedural corpora are compared.

Although Mittani rulers composed letters in the Akkadian language when communicating with other states during the fourteenth century, their vernacular was a language called Hurrian. Neither Indo-European nor Semitic, most of our knowledge of the Hurrian language comes from attestations of personal names and technical terms. Adding to the linguistic medley which characterized the Hurro-Mittanian territorial state was a lexicon that included Indo-Aryan terms, presumably introduced to the region sometime in the sixteenth century. The term for "horse-trainer," or Indo-Aryan ^{LÚ}*assussanni*, was one of these terms. Indo-Aryan *technici termini* related to horses also described their various colors; such terms attested at the site of Nuzi in northern Iraq, which describe the coloration of horses, include the Indo-Aryan *p/babru-nnu* for "reddish brown" and *p/barita-nnu* meaning "grey."[12] These philological details, exhaustive as they may seem, point to an important difference in the ways scribes envisaged technological knowledge: whereas today, technology indexes an imagined future, for cuneiform scribes technology was conceived with relation to a distant past. Hittite technical procedure for training horses reflected both the importance of equestrian knowledge to contemporary communication and warfare, as well as a scribal process of linguistic archaeology.

We see this process most clearly in the Kikkuli text (Figure 6.1), which preserved rare Indo-Aryan numerical terms that indicate the number of laps or "rounds" *wartanna* (from the Vedic *vartanī*) that horses were to run throughout

FIGURE 6.1. The Kikkuli Tablet. Source: VAT 6693 ©Staatliche Museen zu Berlin – Vorderasiatisches Museum. Photo: Sandra Steiß.

the course of their endurance-conditioning. These rare Indo-Aryan numerical terms were: *aikawartanna* "single round," *t(e)riwartanna* "three rounds," *panzawartanna* "five rounds," *šattawartanna* "seven rounds," and *nawartanna* "nine rounds." It is not well understood why the Indo-Aryan terms designated only odd numbers of rounds. Indeed, our present confusion with this has historical precedent, as scribes of the Kikkuli text would supply both Hurrian and Hittite terms to account for even numbers of rounds.[13]

To get a sense of how the Kikkuli text represented these ancient numerical terms, we may turn to the fourth tablet of this five-tablet series. Here, the scribe employs our Indo-Aryan number term *t(e)riwartanna* "three rounds":

> In the morning (the horses) are hitched up and (the trainer) drives them for one-half "mile" and twenty "field lengths." Then he gallops them for one-half "mile" and seven "field lengths" (equivalent to three rounds)—*triwartanna*. In addition, they return to the city at a gallop. When they are unhitched, then they are covered and brought into the stable.
>
> After a while the stable becomes very warm. When the horses become restless and begin to sweat, (the trainer) removes their halters and blankets. Then he puts on their reins and they are led out of the stable. They are bathed in warm water. Furthermore, they are led down to the river and splashed four(?) times with water. Each time they are given one handful of hay and one handful of barley. When they come up from the water, they are led into the stable and now given one quart of groats. Afterwards their barley ration is set out.[14]

A few metrological terms require clarification. First, "miles" indicates the unit of measure known as the DANNA; equally, a "field length" indicates an IKU, a subunit of the DANNA. Second, and notably, whereas IKU in both Akkadian and Hittite sources is traditionally a unit of area measure, in the horse-training procedures it functions exclusively as a unit of linear distance. How far were the horses to run according to tablet IV of the Kikkuli text? The answer to this question is a matter of some debate.

The IKU represents a 1/100th subunit of the DANNA; thus, half a DANNA is 50 "field lengths" (IKU); accordingly, a *triwartanna* is defined here as ≈57 "field lengths" (1/2 DANNA+7 IKU). There are two general schools of thought regarding the modern distances covered by the horses in these texts. According to the first interpretation, the IKU is a short distance equivalent to 15 meters, making the DANNA 1,500 meters.[15] Taking this hypothesis into account, the procedure above recounts an initial distance of a 1,050 meters gallop followed by a *triwartanna* of ≈855 meters. In this case, a day's work measures just under 2 kilometers of chariot sprints (not accounting for the distance needed to travel to and from the city). These figures are noted with utmost caution, however, as the modern equivalencies for the DANNA and IKU remain a matter of significant contention among specialists. As noted, there are two general interpretations for the distances covered. The second, states that the DANNA is much closer to 10.8 km in modern terms.[16] If the short distance for the DANNA is employed, then we can conceptualize the Indo-Aryan "rounds" as sprints designed to acclimatize the horse for the quick bursts of speed needed in chariot warfare. A longer 10.8 km DANNA, in contrast, would significantly change the distances covered more than tenfold. What would not change, however, is the cultural and historical significance of these linear measures. Regardless of whether we assign short- or long-distance equivalencies to the DANNA and IKU, the significance of the *triwartanna* and other terms like it, is that even by the thirteenth century, it was deemed precise, and moreover, part of an ancient—and thereby authoritative—metrological system particular to horse-training.

Whereas in the Hittite procedure above the grammatical subject is in the third person (i.e., the horse-trainer), the Assyrian procedures (composed in the Akkadian language) are directed to the now familiar second-person grammatical subject, the generic "you," well represented in OB mathematical contexts. Much like the second part of the Hittite procedure above, the Akkadian texts are concerned with horse care, typified by instructions such as the following:

> You rub down the horses with warm water, they go down to the water, and as soon as you let them go down, you show them the water (so that they will not shy), you rub them down (again) and release them.[17]

Unlike the Hittite instructions, the sections of the Assyrian tablets that would relate the knowledge to a particular Mittanian horse-trainer are unfortunately broken.[18] Nevertheless, the concern with linear distances and with technical terms related to the care of horses is extensive and well attested throughout the Assyrian procedures. Even where the texts are fragmentary (ellipses below indicate tablet breaks), the Assyrian texts retain much of the same type of information we encounter in the Hittite texts, and as we find in the excerpt below, the sections indicating distances are repetitive:

> They (the horses) will run 2 1/2 "miles," one half (of that distance covered) with a *tahapšu*-blanket . . . You unyoke them, and you rub them down with warm water . . . you harness and tie them up, and they run 2 1/2 "miles," one half (of that distance covered) with a *tahapšu*-blanket . . . for 7 "field lengths" their reins should be taut . . . They run 2 1/2 "miles," one half (of that distance covered) with a *tahapšu*-blanket and for 7 "field lengths" their reins should be taut.[19]

As with the Kikkuli text, the distances covered are measured in both "miles" (DANNA, here, written *bēr eqli*) and "field lengths" (IKU, written *ikê eqla*). Although no Indo-Aryan numerical terminology appears in the Assyrian procedures, ties to Hurro-Mittanian lexicography are found throughout the corpus. Among these, the term *tahapšu*, presumably a Hurrian meaning "clumping," which enters the Assyrian lexicon as a type of felt, i.e., a felt "blanket" as translated above. In codifying knowledge of horse-training and conditioning, scribes established *termini technici* that included numbers, distances, tools, and procedures for maintaining the horses' health. While codifying a particular conditioning procedure for horse-training may seem worlds apart from the earlier OB mathematical problem texts, it is important to note here that the consequences of deviating from the order of operations that called for specific periods of hydration could result in equine colic, and almost certain death for the animal. Moreover, while the "field lengths" of the horse-training procedures referred to specific distances and their subdivisions, they may also be understood as representing set units of time, an always critical component of technical knowledge transmission, as we shall find below.

For Assyrian scribes, there were good reasons to codify and preserve equestrian procedural knowledge at the state capital. Horses were essential for two purposes at Assur. The first, and most well known, is related to chariot warfare. The second, and perhaps lesser-known purpose for equestrian training knowledge was to improve the speed of communication. Stephan Jakob has discussed that horse messengers could travel up to thirty-five miles a day (2003: 291ff). Provisions for horses were detailed when state officials visited neighboring territories (ibid., 2009: 57). Moreover, territories conquered by Assyrian armies, like Nairi (an area to the north of the Taurus Mountains), were targeted specifically for their equestrian knowledge. Indeed, it would not be hyperbolic to state that during the thirteenth century BCE, the horse became one of the key instruments of Assyrian territorial expansion. Even if the horse-training procedures were not used by Assyrian trainers in the thirteenth century, they were still of tremendous value as artifacts of what might be thought of as "original" knowledge, i.e., documentation of specialized techniques from a historical era. It is important to recall, moreover, that the Mittani were once a martial enemy of both Hittites and Assyrians. Thus, with the collapse of the Mittani territorial state, and the subsequent division of the northern Syrian Habur territory situated between Hatti and Assyria, these technical instructions would have also functioned as scholarly symbols of state power.

The Kikkuli text is not simply a set of stable notes that provides direct practical evidence of Indo-Aryan horse-training practices from the sixteenth century. Rather, the text underscores the degree to which even in thirteenth century Hatti, it was deemed important for scholars to record—and thereby rediscover—technological knowledge from an earlier era. This transmission included the Indo-Aryan number terms, whose preservation served two clear purposes. On the one hand, the Indo-Aryan "rounds" functioned as both linguistic and cultural markers of Hurro-Mittanian identity. On the other hand, these same number terms fossilized a historical system of metrology particular to horse-training, not just for Mittanians, but also for Hittites scribes who later saw fit to record the terms on tablets centuries after the linguistic communities which first produced them had collapsed. Compellingly, we will find a similar phenomenon at play a millennium later, in our final case study, we will encounter a mathematical procedure that preserves Middle Babylonian (*c.* thirteenth-century BCE) metrological data in a third-century BCE context.

Cuneiform horse-training procedures demonstrate the extent to which an Old Babylonian mathematical text format could be adapted and expanded well beyond the thematic, geographical, linguistic, and cultural boundaries of Babylonia. During the late second millennium BCE, procedural texts remained the preferred textual vehicle within which scribes could preserve historical quantities, measures, techniques, and *termini technici*. The late-second-millennium horse-training procedures facilitate a return to one of this chapter's core questions, namely: What is "technology" to cuneiform scribes? Here, while the subject matter is horse-training and conditioning, the textual framework and content of the procedures is decidedly antiquarian. Cuneiform horse-training procedures were not presented as manuals for the benefit of contemporaneous Hittite and Assyrian horse-trainers and armies. Rather, the texts, all situated in state archives, codified lexical and numerical knowledge from a distant past. It bears emphasis that during a period of territorial expansion among competing state powers, it was all the more imperative to record and preserve technical expertise that could change the course of statecraft. Seen in this way, the texts may have also served an arithmetic function; namely, to calculate and measure distances and communication speed at a time when such data would have been critical to military expansion. It is precisely the interface of technology, lexicography, numeracy, statecraft, and antiquarianism that we will explore again, centuries later, within the context of seventh-century Nineveh, the final resting place of a corpus of Neo-Assyrian glassmaking recipes.

TIMEKEEPING AND MEASURE: NEO-ASSYRIAN GLASSMAKING PROCEDURES (7TH CENTURY BCE)

Šumma zagindurâ ana epēšīka "if you want to make lapis lazuli (blue glass)." Thus begins an introductory section from a Neo-Assyrian procedures for making colored glass, read literally, making colored "stones" (*abnu*), as there is no independent term for glass in the Akkadian language. These procedures—which revitalize the procedural phrase common to OB Eshunna, *ana epēšika* "for your making/computing"—can be contextualized within a larger corpus of Assyrian scholarly compendia compiling lexical and scientific knowledge of the material world.[20] Indeed, when the glassmaking tablets from the libraries at Nineveh were first cataloged in late nineteenth and early twentieth century, they were identified as medical texts with lists of stones. Over time, scholars like Reginald Thompson, Heinrich Zimmern, and later, A. Leo Oppenheim discovered additional texts that formed part of a scholarly series entitled "the doors of the kiln" (Oppenheim et. al. 1970). Assyrian scribes compiled these procedures during the reign of Ashurbanipal and dedicated the texts to the king's palace library, and—in one elaborate colophon—to the temple of Nabu, the god of scribal craft.[21]

The Neo-Assyrian glassmaking procedures exemplify the interaction of technology, cuneiform scholarship, mensuration, and timekeeping. First, the glassmaking recipes may be read as chemical instructions offering precise quantities for combining silicates, plant matter, and oxides for the production of colored glass. Furthermore, they encode various timekeeping strategies, including hemerological data for selecting a favorable time to create glass, and the use of "short" timekeeping units such as the double hour or *bēru*;[22] this practical or chemical interpretation of the texts has been favored in Assyriology. Second, the texts also functioned as scholarly investigations of the material world. Unlike earlier glassmaking ingredient lists from the second millennium, these seventh-century texts were scholarly compilations that included a ritual introduction with a kiln-making procedure duplicated on three tablets that invokes purifying deities called *Kūbu*. This same introductory ritual provides an exemplary model of scribal hermeneutics, wherein the polyvalent potential of the cuneiform script is on full display. As such, the glassmaking recipes from Nineveh provide a clear case study for the interaction of knowing-how and knowing-that, *technē* and *epistēmē*. The sections of the glassmaking procedures explored below place particular attention on notions of numeracy, mensuration, and timekeeping within chemical craft contexts.

We may begin with a single exemplary tablet from the corpus at Nineveh. K.4266+ offers a compendium of procedures and ingredient lists for producing a variety of kiln-made precious stones (i.e., colored glass); on the obverse of the tablet we find a recipe for making "lapis lazuli," *uqnû*-colored glass. This is done by combining silicates, and ground plant matter and vitreous materials that functioned as fluxes and coloring agents:

> To 1 mina of [good-quality] *tērsītu*, you mix, repeatedly, and grind: [1/3] mina of ground *būṣu*-glass, 1/3 mina [of *amnakku*], and 5 kisal of (sea) shells. You collect (these ingredients) in a (casting)-mold and close it. You set (the mold) between the openings (of the kiln) using *tašnû*-tongs; that which emerges (is): fine quality "lapis lazuli" (*uqnû*-glass).

> You fire (the following) for seven days within an *utūnu*-kiln. For 1 mina of *tērsītu*: You mix repeatedly and grind 3 minas of *tarabānu*-plant of *šaddu*-quality, 10 kisal of *šaddu*-quality *būṣu*-glass, 5 kisal of carnelian-colored sea shells, 2 kisal of *anzahhu*-glass, 3 shekels of red (*šaršerru*)-paste, 5 kisal of red ["male"] alum, 5 kisal of yellow (*šīpu*)-paste, and 6 shekels of "white plant." You collect [(these ingredients) in a mold]; that which emerges (is): [fine-quality] "lapis lazuli" (*uqnû*-colored glass).

> That which does not "sink" (i.e, melt within the mold), you re-fire for (another) seven days within an *utūnu*-kiln.[23]

Each distinctive section, demarcated on the clay tablet and indicated by the line ruling above, marks progressive steps in the recipe. Much like an algorithm, the glass recipe above offers a linear sequence which is interrupted by a decision-making protocol. Here, in the penultimate section, we are told that the vitreous mixture is to be annealed in a kiln for 7 days to produce a striated "lapis lazuli" colored glass. If, however, the mixture does not melt and anneal properly within the kiln after 7 days, the final section provides an instruction that asks "you" to repeat the 7-day annealing process until the desired result is achieved. This "algorithmic form," as Jim Ritter has argued, is a defining feature of Akkadian procedures introduced historically during the Old Babylonian period:

> [T]he same marking of information flow and use of verbal chains to demarcate different operational levels and their interaction was applied across a broad spectrum of intellectual domains, a spectrum which increased with time as certain practices—astronomy, glassmaking—were found to be reducible to the same *algorithmic form*. (Ritter 2009: 101, emphasis added)

Glassmaking procedures are not always clear as to sequence and timing of all operations. Nor do scribes ever clarify the intended purpose of the glass products described in the texts. This phenomenon is not unusual in procedural texts and recipes, which regularly exhibit a tension between explicit protocols and what sociologists like Harry Collins have called "tacit knowledge" (Collins 2010). Indeed, as Matthieu Ossendrijver has argued within Late Babylonian astronomical contexts, procedural texts are inherently "incomplete":

> As verbal representations of algorithms, procedure texts are often, strictly speaking, incomplete, in the sense that certain operations or bits of information are omitted or formulated ambiguously. Usually the combined analysis of procedure texts and tabular texts enables one to fill out this tacit knowledge and reconstruct the more or less complete sequence of operations that constitute the underlying algorithm. (Ossendrijver 2016: 159)

Another distinctive component of the glassmaking procedure above concerns metrology: specifically, a unit of weight called the kisal. Once thought to be a subdivision of the standard shekel (8 1/3 grams), an unpublished Assyrian weight now indicates that one kisal corresponds to approximately 9.6 grams, as Nils Heeßel has characterized it, "only insignificantly heavier than one sheqel."[24]

Presuming the difference between the standard shekel and the kisal is ≈1.3 grams, we might ask, following Heeßel, was this miniscule weight difference considered truly insignificant to the scribes of the glassmaking recipes? A clue may lie in a curious Akkadian expression preserved on the reverse of the same tablet; it states: "(These are) the ingredients, [not (missing even a single)] 'hair of [barley].'" The phrase appears to underscore the degree to which metrological precision was not only necessary, but essential, suggesting that if even a single gram were missing, the procedures would fail. Interestingly, the reverse of K.4266+ also compiles pithy sets of instructions for reproducing various types of "stones"; these instructions forego the usual second-person procedural format, and are akin to earlier second-millennium ingredient lists, supplying only quantities and ingredients for glassmaking:

3 minas of *tērsītu*,[25] 8 minas of (. . .) *zukû*-glass, 1 1/2 minas of alum, and 12 kisal of [anzahhu-glass]: (these) are the ingredients for (. . . quality) "lapis lazuli" (*uqnû*-glass).

---

1 mina of *tērsītu* and 2 minas of *šadû*: (these) are the ingredients for fine-quality "lapis lazuli" (*uqnû*-glass).

---

1 mina of *tērsītu* and (. . .): (these) are the ingredients for (. . .) of "lapis lazuli" (*uqnû*-glass).

---

1 mina of *zukû*-glass, 15 kisal of (. . .), 10 kisal of lead: (these) are the ingredients for Elamite carnelian according to (the expert) *Wa*-(. . .)

---

1 mina of white "quartz," 15 kisal of *tuskû*: (these) are the ingredients for Assyrian carnelian.

---

1 mina of white "quartz," 15 kisal of *tuskû*: (these) are the ingredients for Assyrian alabaster.

In addition to the metrological minutiae detailed above, the texts preserve clear instances of ritual and practical timekeeping, indicating a link to the astral sciences. Glassmaking procedures on tablet K.4266+, for example, are interjected by a ritual that calls for sacrificing sheep and setting up purifying deities called *Kūbu*, (a similar ritual is attested in an introductory section of the recipes). Here, we are given instructions for setting up these deities within a specified time frame:

> In the process, you set up *Kūbu* within two double hours (*bēru*). You sacrifice a sheep. You make a funerary offering to experts of yesteryear. You collect the ingredients in a (casting)-mold and set it down into an *utūnu*-kiln

The "double hour" or *bēru*, is a technical term we have come across before in the context of horse-training procedures. There, the *bēru* (written DANNA) referred to a linear distance, the so-called "mile" of ≈10.8 km. In the glassmaking recipes, however, *bēru* takes its second meaning, as a fixed-length time unit of 2 hours (a "double hour" in the translation above); a subdivision of a full day (sunset to sunset) consisted of 12 *bēru*. As John Steele has recently summarized, there is reason to believe that the two meanings of *bēru* are historically linked, stating that "it seems highly likely that the *bēru* of distance corresponds to the distance walked in one *bēru* of time" (Steele 2019b: 97–8).

In first-millennium astronomical contexts, *bēru* measurements were employed relative to a reoccurring phenomenon, as in, for example, 4 *bēru* before sunrise or sunset. Particularly in recording time intervals, such as in water clocks and sundials, the *bēru* is among the most consistent subdivisions of "short" time measurement practices we find evidence for in the ancient world. There is further evidence that the *bēru* could correspond to a weight measure; for example, 1 *mina* could be equivalent to 1 2/3 *bēru* (or 3 hours and 20 minutes) within the context of an idealized water clock that tracks and measures the weight of water relative to the passage of time.[26]

Time is represented and codified in three distinct ways within the glass recipes. First, we find instructions that indicate one is to purify the glassmaking "laboratory" within an ideal time frame of ≈4 hrs, i.e., 2 *bēru*. Second, the recipes record a specific length of days for annealing kiln-made vitreous materials. On the matter of annealing glass, the recipes shift between seven- to ten-day intervals when providing instructions for making glass in a kiln, roughly corresponding with the modern recommendation of ≈250hrs annealing time kiln-made glass. One instruction states, for example, "that which does not 'sink' (i.e, melt within the mold), you re-fire for (another) seven days within an *utūnu*-kiln." Third, and finally, these recipes designate a favorable or lucky time during which glass should be made, echoing a well-attested calendrical and scholarly practice known as hemerology.

In the instructions that employ a bronze alloy for coloring glass called "fast bronze," for example, we find instructions for fritting vitreous materials on a "favorable" day:

> If you want to make "fast bronze," you (first) set down, into a cold kiln, (. . .) minas of salicornia and 10 minas of *amnakku*-sand; you ignite a proper fire. You frit (the ingredients; lit: "you cause them to turn white"). You frit them once again, using an old *harāgu*-crucible. You pour [water onto the mixture] (and) on a favorable day you make an offering at the opening (of the kiln). You set up a censer of juniper.

Finding the right time of month within which to produce glass, or during which to construct a four-chambered glassmaking kiln, was dependent on having knowledge of the scholarly practice of hemerology. A curious feature of the glassmaking procedures is that while they employ a hemerological framework, contemporaneous compendia of Assyrian hemerologies make no mention of lucky days for glassmaking. Here, there is good reason to suspect that what we are reading may best be characterized as a "mock" hemerology, a concept best explained by examining the excerpt below:

> ***qurū***-wood that have no knots on the surface, bound up with an ***apu*** "leather strap" that you cut in the month of *Abu*, these logs should go into the bottom of the kiln (*kūru*)[27]

The two Akkadian terms italicized in bold above are hapaxes, i.e., lexemes which appear exclusively in this corpus of technical procedures. However, they do not exist in isolation. When contextualized within a scribal context—recall that the Neo-Assyrian glassmaking recipes are dedicated to the gods of scribal knowledge—it is revealed that the seemingly hemerological passage above is better understood as an exercise in scribal hermeneutics. As I have argued elsewhere, the month *Abu* (August/September), written using the cuneiform sign IZI (the Sumerian term for "fire"), is a homophone of the hapax *apu* "leather strap," which may also be read *abu* or "father" (Escobar 2019). What appears to be a calendrical record, is rather a multivalent passage that conceals multiple levels of meaning, as demonstrated by a comparative translation of the passage in question, which juxtaposes a standard translation (left) of the instruction with a richer, scribal translation of the passage (right) that attempts to capture how erudite Neo-Assyrian scribes employed the cuneiform script to generate new semantic and technological possibilities:

Standard Akkadian translation	**Multivalent scribal translation**
. . . *qurū*-wood that have no knots on the surface, bound up with an *apu* "leather strap" that you cut in the month of *Abu*, these logs should go into the bottom of the kiln (*kūru*).	"kiln"-wood (*qurū*) logs with no knots on the surface, bound up with the "father"-strap (*abu*), the logs that you cut in the "month of fire"—also called the month of the father (*Abu*)—these are the logs that should go into the bottom of the kiln.

Neo-Assyrian glassmaking procedures demonstrate the degree to which what may appear as solely a "practical" manual can encode scholarly and theoretical knowledge. As scribal reflections of glassmaking technology, these procedures were carefully curated to include and highlight that which was deemed most important for the scholarly libraries at Nineveh. As such, the modern translator may be unable to determine whether a given procedure was meant to reflect an anthropological record of true craft practice, or else, whether in certain instances procedure texts could veer closer to hermeneutical exercises, "mock" representations of actual practice. The concept of a mock glassmaking procedure—or hemerology as in the passage above—may indeed have mathematical analogues. Jens Høyrup, in examining an Old Babylonian volume capacity procedure from Eshnunna and its Egyptian counterpart, evokes the sixteenth-century concept of a *Schimpfrechnung*, or "mock reckoning," which he defines as:

> a challenge meant to impress and make fools of the non-initiate. Problems of this type turn up regularly in medieval and Renaissance treatises on applied mathematics that draw regularly on oral or semi-oral practitioners' traditions—exactly the traditions where rules for practical computation go together with mathematical riddles that seem to refer to practice but rarely have any practical application. (Høyrup 2002: 321)

Cuneiform procedures provide an ideal textual format for encoding scribal knowledge within the linguistic framework of a practical manual. The cuneiform writing system itself, wherein one sign could mean many things, served only to amplify the hermeneutical interface of theory and practice, a relation of knowledge types that we will examine with one final case study from the Late Babylonian period.

CALCULATING THE DIMENSIONS OF THE "TOWER OF BABEL": THE ESAGILA TABLET

Procedural texts witnessed a reemergence of metrological and mathematical subject matters during the latter half of the first millennium BCE and throughout the Late Babylonian period. Lacking a native imperial capital, Late Babylonian scribes turned to city temples and private households, rather than state archives. During the late periods of cuneiform scholarship, the scholarly landscape was dominated by cities like Babylon and Uruk, where elite scribal families copied and generated procedural texts encompassing a range of specialized genres, from mathematics and astronomy to medical recipes.

Among the Late Babylonian procedures that have gained broad attention is a tablet found at the Bīt Rēš temple of Anu at Uruk, known as "The Esagila

FIGURE 6.2. The Esagila Tablet *c*. 229 BCE. Photo by Wikimedia Commons.

Tablet" (Figure 6.2). The text—which dates to 229 BCE (SE 83), and purports to be a copy of an earlier original—provides instructions for surveying the dimensions of sacred structures like the E-sagil temple of Marduk in Babylon and the Etemenanki, a ziggurat which has been linked in literary traditions to the Tower of Babel (George 1992: 109–19). The procedures compiled on the tablet function as mathematical exercises, and provide instructions for converting area dimensions.

Calculations for the base of the Entemenanki ziggurat are explored in two related ways. First, (in lines 16–19) the scribe provides the dimensions of the base in linear and surface "reed measures" and converts these to capacity "seed measures" using Neo-Babylonian metrological units. In the section that follows (lines 20–4), the same conversion is done using a more ancient metrological system dating to the Middle Babylonian period. The calculation below begins with the standard procedural phrase "for you to calculate/see" *ana amarīka*, a variation of which, we should recall, was employed in the earlier glassmaking and mathematical procedures:

> obv. 16–19. For [you] to see the measurements of the base of (the ziggurat) Etemenanki, length and width. The length is 3 chains (*ṣuppān*), the width 3 chains (*ṣuppān*), by the cable-cubit (*aslu*). To multiply their calculation, 3 [times 3] is 9 : 9 times 2 is 18 (*sūtu*). If you do not understand 18 (*sūtu*), it is 3 (*pānu*) seed (measure) by the small cubit. The base of Etemenanki. The height is in accordance with the length [and width].[28]

The passage begins by multiplying the length and width of the structure using *ṣuppān* (approx. 30 m), which is based on the cable cubit or *aslu* (approx. 50 cm). Multiplying the length and width yields 9 *ṣuppān* squared or approximately 8,100m^2. This result is then converted into the *arû* "seed measure" used for calculating grain capacity, 9 *ṣuppān* is equivalent to 18 *sūtu*; 18 *sūtu* in turn is equal to 3 *pānu*, clarifying correspondences between the two metrological systems. The base of the Entemenanki ziggurat is represented as an idealized cube structure, where the height, length, and width are equal.

A fascinating feature of these procedures is that, much like the earlier Assyrian and Hittite texts, the doer/scribe is provided with an alternative instruction for arriving at the same result. In the passage below, what is offered is another method for measuring the base of the Etemenanki using an older *arû* cubit measure (approx. 75 cm). By the Late Babylonian period, the *arû* cubit was already an archaic system which had long fallen out of use. Mirroring the preceding passage, these ancient dimensions are offered in both linear and capacity measures conversions:

> obv. 20–24. Otherwise, for [you] to see the measurements of the base of Etemenanki, length and width. The length is [10 rods], the width 10 rods by

> the *arû* cubit. In order to multiply its calculation, [10 times 10 is] 1 40 : 1 40 times 0; 18 : 30. If you do not understand 30, it is 30, where 1 (*ikû*) area is 5 (*sūtu*) seed (measure) [by the] large [cubit]. The base of Etemenanki. Length, width, and height are 10 rods square [by] the *arû* cubit. (Robson 2007a: 177–9)

In juxtaposing historical metrologies, the Esagila Tablet lends support to a scholarly interpretation of technical procedures. As Andrew George has argued, the tablet "presents an ideal of how the tower was meant to look, and does not constitute a physical survey of a built structure." George adds that the various features linear area and capacity conversions calculated on the tablet, as discussed above, are "more abstract than architect's plan. The suspicion must be that as an extrapolation of arithmetical and geometric problems taking as its subject idealized dimensions of Marduk's temple precinct and ziqqurat, the text of the E-sangil Tablet originated a compilation of training exercises in mathematics for would-be surveyors" (George 2005/2006: 77).

In providing instructions for what appear to be idealized dimensions of the sacred building's foundation, these arithmetical instructions reconcile two historical systems of metrology. However, the E-sagil Tablet is much more than a simple conversion exercise. As a scholarly document, the text creates an arithmetic dialogue with surveyors concerning a deep Babylonian past, a dialectical connection made all the more emphatic by the grammatical subject of the text: "you." Only with proper knowledge of the history of both the Etemenanki and Babylonian area metrology can "you" calculate (literally, "see," Akk. *amāru*) the dimensions of the sacred structure.

As with many scholarly texts, the Esagila Tablet ends with a colophon indicating the hermetic nature of the knowledge contained therein: "The knowing may show the knowing; the unknowing shall not see it. Written, made good, and checked against an old tablet from Borsippa." Here we may pause on yet another conceptual difference between cuneiform scholarly procedures and contemporary instructions or recipes today. Rather than aim to transmit knowledge from expert to novice or from "knower" to "non-knower," (in Akkadian, *mūdû* to *lā mūdû*)—what we find in cuneiform procedures is an expert-to-expert model of transmission—practical and mathematical knowledge encoded by scribes exercising their skills within enclosed scholarly communities.

CONCLUSIONS: PROCEDURAL KNOWLEDGE IN TECHNOLOGICAL AND MATHEMATICAL TEXTS

This chapter has presented evidence for two related conclusions. The first is that for cuneiform scribes, technology was not understood in terms of "applied science," nor was it defined in terms of an opposition to scholarly inquiry. On the contrary, cuneiform procedural texts provide abundant evidence for an intellectual history technology. Second, a historical survey of instructional texts in cuneiform scholarship demonstrates that technical procedures in Mesopotamia were historically rooted within a Babylonian mathematical problem text tradition that persisted for two millennia. Adapting to their local cultural contexts, cuneiform procedural texts—the instruction manual, understood as a scientific object in its own right—would prove to be among the most successful vehicles for the textual transmission of technical knowledge in the ancient Middle East. The texts we have examined have focused specifically on how technical subject matters like horse-training, glassmaking, and architectural surveying, were reshaped according to the scholarly needs of distinctive scribal communities. Having examined where and how technological knowledge is manifested in scribal cultures, we can now reconsider how procedures may reshape our intuited notions of expertise, the transmission of technical knowledge, and the fundamental co-dependence of making and calculating in cuneiform cultures.

Furthermore, Akkadian technical procedures provide evidence for scribal and cultural notions of technical and scientific expertise that predate—and by extension, circumvent—many of the traditional knowledge hierarchies distinguishing between making and theorizing. The semantic link in Akkadian between making and calculating lies at the heart of cuneiform procedural knowledge. This same coupling of head and hand is extant in Akkadian actor terms. Take, for example, the term *ummânu*; it is the Akkadian noun that means "expert," but can refer simultaneously to a skillful scribe or a knowledgeable glassmaker, an expert mason, horse-trainer, or a scribe versed in mathematical astronomy. An *ummânu* could, in a modern sense, refer to both a mathematician or an artisan without any obvious epistemic value distinction. A section of a seventh-century BCE Assyrian glassmaking recipe which we have already examined, makes vivid this double semantic sense of *ummânu* in the Akkadian language:

> In the process, you set up *Kubu* within two double hours. You sacrifice a sheep. You make a funerary offering to the experts (*ummânu*) of yesteryear. You collect the ingredients in a (casting)-mold and set it down into an *utūnu*-kiln

Who were these *ummânu*-experts for whom animal sacrifices were to be made? Were these *ummânu* glassmakers or scribes? Technicians or literati? Translating the phrase *ana ummânī kispa takassip*, "you make a funerary offering to the experts," furnishes both modern and ancient notions of technical expertise. Here, the *ummânu*-expert is one who has knowledge not just of glassmaking, material properties, and kiln casting, but also of ritual purification, and astronomical timekeeping (the "double hour" or *bēru*, as discussed above, is common in astronomical contexts). Demarcating a strong boundary between knowing-how and knowing-that would be at odds with how the more holistic frameworks of *ummânutu*, "expertise," *nēmequ* "wisdom and knowledge," and *ṭupšarrutu* "scribal craft." Equally, in Akkadian, someone who is "skilled" or "wise" is *emqu*, a derivation of the verbal roots (*nmq*), *nēmequ*—simply, knowledge.

Translating and comparing modern terms for expertise, calculation, and craft against their Akkadian counterparts reveals the extent to which the familiar tension of theory and practice initiated in Classical cultures is not clearly reflected in cuneiform scholarship. For cuneiform scribes, the transmission of "technology" was an exercise in linguistic archaeology, bound to a procedural text category that could trace its historical roots to a mathematical text tradition. Employing an algorithmic style, procedures preserved decision-making protocols, recorded precise quantities, reconciled historical metrologies, codified and generated numerical and technical terms, and provided instructions for timekeeping. All of this, scribes did within local contextual frameworks of state and private archives by adapting an ancient textual framework rooted in Old Babylonian mathematics to the contexts of their own time. In reflecting on the cultural longevity of cuneiform technical procedures, an analogous intellectual history is worth considering, namely, technical and procedural knowledge within the Sanskrit tradition of *śāstra*. As Sheldon Pollock relates:

> All knowledge derives from *śāstra*; success in astrology or in the training of horses and elephants, no less than in language use and social intercourse, is achieved only because the rules governing these practices have percolated down to the practitioners—not because they were discovered independently through the creative power of practical consciousness—"however far removed" from the practitioners the *śāstra* may be. As for learning the *śāstra* itself, this is the necessary commencement of the tradition, and later serves to enhance the efficacy of the practice . . . [i]t makes the practice "stronger," reliable, and consistent, unlike uninformed practice, whose effectivity is altogether fortuitous. (Pollock 1985: 507)

Situated beyond the binarism of theory vs. practice, or *epistēmē* vs. *technē*, what we encounter in the cuneiform technical procedures is, to employ Pollock's terminology, a co-dependent *strengthening* of mathematical and technological knowledge transmission. Rather than define technology derivatively as "applied science," cuneiform technical procedures serve as evidence of a long-standing scholarly inquiry into the transmission of practice, calculation, mensuration, lexicography, and material transformation. Herein cultural knowledge of volume, weight, distance, area mensuration, glassmaking, horse-conditioning, and history itself could be codified procedurally, thereby strengthening technical practices and scribal arts in tandem.

CHAPTER SEVEN

Representing Mathematics

Depictions of Ancient Mathematics in Word and Image

LIBA TAUB

The mathematician Euclid of Alexandria (*fl. c.* 300 BCE) was famously reported—by Proclus (410 or 412–485 CE), a much later commentator—to have replied to a question posed by the Pharoah Ptolemy I Soter about whether there was any shortcut to learning geometry: there is no royal path. Mathematics has often been presented as difficult, and yet it held an unmistakable appeal for some. The Athenian philosopher Aristotle (384–323 BCE) had recognized that different people require different styles of communication and that certain audiences have clear expectations: some people will not accept what a speaker says without a mathematical proof; others require illustrations, while still others want poetry. As we will see, a number of mathematicians—including the renowned Archimedes—were credited with having composed mathematical problem-poems, while the revered epic poets, Homer and Hesiod, were celebrated as expert calculators.

Mathematics was part of broader culture and society in antiquity, not confined only to specialists. Ancient written and visual depictions of mathematics—including activities involving mathematical skills, as well as individuals identified as mathematicians—are explored here, with a focus on Greek and Roman culture. Mathematics was represented as having different uses: practical (including measuring and calculating) and theoretical (think here about Plato, Aristotle, Euclid, and Ptolemy), as well as recreational (involving riddling problems, and games, such as dice). There is evidence that a good number of people, mainly but not only men, took pleasure in displaying—even flaunting—their mathematical knowledge. The practical desirability of certain calculations (related to concerns about land boundaries and taxes) relied upon the special intellectual ability and training required by such tasks. Some notable accounts of mathematical problems and solutions were—literally—legendary (think of Archimedes in the bath).

DEFINING MATHEMATICS

The English word "mathematics" tends to have a narrower focus than the ancient Greek word from which is derives. As we saw in the Introduction, *mathēma* (plural = *mathēmata*) was used to refer to "that which is learnt," "learning or knowledge," as well as the "mathematical sciences." The word "mathematics" can be understood as "things to be learned"; the Greek noun *mathēma* derives from the verb *manthanō*, "I learn." The Greek word *mathēmata* is often understood as referring to learning, more generally than just the "mathematics" implied by the English term.[1]

In antiquity, there was not total agreement about what mathematics comprised. A number of later Greek authors—including Porphyry (234–*c.* 305 CE)—claimed to quote from the Pythagorean Archytas of Tarentum (*fl. c.* 400–350 BCE), writing about the *mathēmata*, referring to geometry, arithmetic, astronomy, and music as being as kin, or siblings, to one another. Plato (*c.* 429–347 BCE), who knew Archytas personally, also referred to the kinship of astronomy and harmonics. Not all ancient authors agreed as to what should be included in a definition of mathematics. Proclus relates that four kinds of mathematics had been identified by the Pythagoreans, but he also noted that Geminus (probably *c.* 50 BCE) divvied up mathematics rather differently.

Some branches of mathematics have names that carry other meanings, referring to physical bodies, such as measurement of the earth or land (geometry), knowledge of the stars (astronomy), and musical theory (or how things are joined together = harmonics). These other meanings color the representation of mathematics, indicating a degree of ambiguity about the nature of the subject. In *Clouds*, a comedy by Aristophanes (second half of 5th century BCE), characters joke about whether geometry (Greek *gē* = earth; *metron* = measure) refers to measuring agricultural land, or the entire earth.

THE ORIGINS OF MATHEMATICS: GIFTS FROM GODS, INVENTIONS OR DISCOVERIES?

Ancient sources in several cultures portray the invention or discovery of mathematics and related skills as being associated with gods or with royalty. Babylonian accounts of mathematics display several features that recur in other cultures, including Egyptian and Greek. For example, mathematics is depicted as divinely enabled. Babylonian sources indicate that literacy, numeracy, and the skills required for linear measurement were considered to be divine gifts from goddesses, who bestowed numeracy and mathematics on humanity to enable rulers and their servants (that is, trained scribes) to deliver social justice through the equitable distribution and fair management of assets. The Sumerian goddess Nisaba is described as providing a reed, a measuring rope of lapis lazuli, a yardstick and a writing board, as well as giving wisdom, which she generously bestows on the king (Robson 2009: 218). Indeed, Babylonian mathematical practices are sometimes linked to royal prerogatives. Studies of Old Babylonian (*c.* 2000–1600 BCE) scribal schools provide glimpses of those engaged in mathematics: small numbers of professionals within a restricted social circle were numerate. Old Babylonian mathematics was concerned with quantity surveying and labor management as well as the abstract manipulation of lines and areas (Høyrup 2018). Mathematics was valued for many practical matters; in some areas, fairness and justice relied on numeracy and metrological skills.

Babylonian mathematical skills were highly regarded elsewhere in the ancient world. Greek writers tended to trace important aspects of their civilization to the influence of Eastern neighbors, and some credited Babylonians and/or Egyptians with the invention of mathematics. Geometry, in particular, is said by a number of Greek authors living in different periods and writing from different perspectives—including Herodotus (*c.* 484–*c.* 425 BCE), Diodorus (first century BCE), Diogenes Laertius (probably first half of the third century CE), and Iamblichus (third century CE)—to have originated in Egypt.

Herodotus thought that the annual flooding of the Nile required surveying skills to determine field boundaries, and that this in turn led to the invention of geometry. Certainly, measuring and calculating were involved. He reports that, as part of his taxation regime, the Egyptian king Sesostris had the land divided into portions, and that this may have been part of the invention of geometry (in this case, literally, land measurement). In Book II of his *Histories*, Herodotus credits the learning of geometry by the Greeks to their knowledge of Egyptian practices, from whom he thought the Greeks had learned how to measure land. However, he emphasized that the sundial, and the twelve divisions of the day, came to Greece from Babylonia and not from Egypt.

Some Egyptians reportedly believed that mathematics had been given to them by the god Thoth. Socrates, a speaker in Plato's *Phaedrus* recounts a myth that the Egyptian god discovered numbers and arithmetic and geometry and astronomy, as well as dice and draughts. In the *Metaphysics*, Aristotle also suggests that Egypt was the birthplace of mathematics, because there the priestly class had enough time for its study. In the *Laws*, Plato admiringly referred to Egyptian children's knowledge of arithmetic, advocating that Athenian children should follow their example, learning mathematics alongside the alphabet.

Texts describing and portraying mathematics and mathematicians were not always written by mathematicians themselves. The Greek tragedian Aeschylus (? 525/4–456/5 BCE), in *Prometheus Bound*, portrayed the Titan Prometheus as proclaiming that he was responsible for teaching humans the civilizing arts, including agriculture, medicine, and metallurgy. Prometheus is portrayed in literature as a tricky character, who stole fire from the gods, but nevertheless benefited humans. He claims responsibility for having discovered (or invented) numbers, the most important of the arts, as well as writing (in the so-called "Catalogue of the Arts"); he was also associated with the development of astronomy and arithmetic. Aeschylus' crediting of Prometheus with mathematical prowess has some similarities to Babylonian and Egyptian accounts of mathematical skills and practices conveyed as gifts from gods and goddesses. Intriguingly, in the *Tusculan Disputations*, the Roman writer Cicero (106–43 BCE) refers to Aeschylus as not only a poet, but also a Pythagorean, perhaps giving insight into an apparent interest in mathematics. (Cicero also describes his own uncovering of the neglected tomb of Archimedes (*c.* 287–212/1 BCE), signaling his respect for Greek mathematics.)

A number of Plato's dialogues vividly portray discussions about mathematics. His dialogue, the *Meno*, may be the earliest extended evidence regarding Greek mathematics. The genre of dialogue, particularly in the Socratic form devised by Plato, is well suited to the presentation and consideration of problems. However, very few of the "problems" presented in the Socratic dialogues are concerned with mathematics; rather, Plato was concerned with philosophical issues. For example, in the *Theaetetus*, the problem of incommensurability is touched upon only briefly, while the question of the nature of knowledge underpins the work as a whole. Nevertheless, Plato has often been understood to have emphasized the value of mathematics, for a variety of purposes, including the creation of the world (as we shall see below) as well as political leadership. Plutarch (before 50–after 120 CE), one of the few ancient authors to compose a dialogue concerned with scientific questions, did present some mathematics in his dialogue *On the Face on the Moon*. Yet, surprisingly, of all the interlocutors named in the dialogue, it is only the one described as a mathematician, Theon, who never himself speaks. Plutarch presents the mathematician as a silent participant in the discussion of the problems posed.

In Plato's *Republic*, Socrates states that the Trojan War hero Palamedes claimed to have invented numbers. He was also credited with inventing weights and measures, dice, the alphabet, and mealtimes (Figure 7.1). The Greek author, Pausanias (*fl. c.* 150 CE), mentions the temple of Fortune in which Palamedes dedicated the dice that he had discovered. Paintings depicting a game of dice being played between the heroes Achilles and Ajax are known on over 125 Greek vases, suggesting the popular appeal of the subject. On some of these paintings, the number of points won, four (*tessera*) and three (*tria*), are indicated by what can be made out from the inscriptions in front of their faces, almost as if coming from their mouths, rather like a modern cartoon. The higher number goes to Achilles, and he is the winner, while Ajax is the loser. While the playing of dice may not have required mathematical skills or knowledge, recognition of and some understanding of numbers was essential to the game. Pausanias also described dice depicted in a painting, in a way that suggested that it was a familiar game: "if you turn your gaze again to the upper part of the painting, you see, next to Actaeon, Ajax of Salamis, and also Palamedes and Thersites playing with dice, the invention of Palamedes." That we have so many vase paintings depicting men playing dice as well as a written description of such a painting, indicates that this game that required a certain level of numeracy was well known and understood.

The game of chess appears to have originated in India during the Gupta Empire (third–sixth centuries CE), been introduced during the Sassanid Empire to Persia and then, after the Islamic conquest of Persia, spread in the Islamic world. While the origin story has various versions in several languages and cultural settings, the core elements are very similar. In brief, the ruler of India witnessed a game of chess at court and was fascinated. The inventor was summoned and congratulated on the invention, and was promised the granting

FIGURE 7.1. Cube and bone dice. National Museum of Roman Art, Merida, Spain. Photo by PHAS/Universal Images Group / Getty Images.

of any wish as a reward for cleverness. In the version recounted by the ninth-century historian Ibn Waḍīḥ Al-Ya'qūbī—one of the oldest written accounts of the story—the inventor (who was a philosopher) asked for a quantity of grain to be counted out on a chessboard:

> one grain for the first field of the board, two grains for the second, and so on, so that each field contained twice as many grains as the one preceding it. The grain stores of that reign were not enough to grant [the inventor's] wish, but he said he did not need such a quantity, adding that for him just a small piece of lawn was enough. The queen then asked him about the number of the grain corns he wished. He therefore explained the computation: the first row has 255, the second one 32768, the third 8388608, the fourth 2147483648, the fifth 549755813888, the sixth 140737488355323, the seventh 36028797018963968, and the eighth 9223372036854975808. The total of the corn grains on the chessboard is equal to 18446744073709551615. (Bardi 2019: 4)

This is the "wheat and chessboard" problem, which relies on a geometric progression, that is, a sequence of numbers in which each term after the first is found by multiplying the previous one by a non-zero number. Geometric progressions were known and used in problems prior to the invention of the chessboard, and were part of recreational mathematics in ancient Babylonian and Egyptian cultures.

Many ancient accounts of the work of mathematicians report their discoveries (things that already exist and are found) rather than inventions (new things that are created). The third-century BCE philosopher Eudemus of Rhodes wrote histories of *mathēmata*; he is considered to be the first historian of Greek mathematics. In his *History of Astronomy*, he presented the contributions of mathematicians and astronomers as discoveries. In fact, this work was referred to in later sources by the title, "Who Discovered What in Mathematics?" That the history of mathematics was seen as a suitable subject for a philosopher's study is an indication that mathematics was regarded as an important form of knowledge. However, in the various accounts of achievements, it is not always clear whether "invention" or "discovery" is meant, whether a contribution is to be regarded as a creative act of invention or as a discovery of something that previously existed. Whether or not mathematics is invented or discovered is still debated by twenty-first-century philosophers.

MAKING PROGRESS

The idea that mathematics is not static and that progress can be made was articulated by the Epicurean philosopher Philodemus (*c.* 110–*c.* 35 BCE). He was reporting an account from an unnamed source, claiming that the philosopher Plato had been the architect of advancement in mathematics, with Eudoxus and his students revising the older (fifth century BCE) ideas of Hippocrates of Chios (Zhmud 1998, and 2020: 273). Later, Claudius Ptolemy, the second-century CE mathematician especially renowned for his work on astronomy, geography, and harmonics, emphasized the idea that progress in mathematics is possible and should be an aim of its practitioners. In the fifth century CE, Proclus reinforced the image of mathematical advancement. He reported that Aristotle had said "somewhere" that mathematics is intrinsically desirable for those who are engaged in it, noting that great progress has been achieved in a short time, without any reward offered to those pursuing it, and that even those who gain just slight benefit are nevertheless fond of it and spend time doing mathematics while neglecting other matters.[2]

With this image of mathematical progress, at least one author argued that allowances should be made for those who worked in earlier periods, when important questions were still unanswered and geometry was in its infancy. Synesius of Cyrene (*c.* 370–413 CE), Bishop of Ptolemais and a student of Hypatia of Alexandria (born *c.* 350–70; d. 415), cited the example of the astrolabe which, building on earlier ideas and methods, represented a later and more extravagant display of mathematical knowledge than was previously available. He shared his views in a letter sent with the gift of an astrolabe to an officer in the imperial court at Constantinople, claiming that everything he knew about astrolabes had been taught to him by Hypatia. (Assuming that this was a planispheric astrolabe, it relied on a mathematical technique known today as stereographic projection.) While expressing gratitude for the work of earlier mathematicians, including Hipparchus, he recognized that their efforts had been surpassed by later endeavors. Synesius regards the astrolabe as a high point in mathematical—and cultural—achievement.

MATHEMATICS AS A SPECIALIST—EVEN LEGENDARY—ENDEAVOR

Mathematics was often portrayed as involving distinctive skills and abilities. The Babylonian text, sometimes referred to by the name "Scribal activities," gives a sense of the skills and tasks mastered by the Babylonian scribes, including mathematics and budgeting. Mathematical practitioners acquired specialist skills, and sometimes used particular tools, such as writing, measuring, and counting implements, including the abacus.

Certain notable Greeks sought to publicize and celebrate their own special mathematical achievements. Archimedes, according to Plutarch, wanted his grave to have a monument portraying a sphere inside a cylinder, a reference to his work *On the Sphere and Cylinder*; he regarded this as one of his greatest mathematical achievements. In his *Life of Marcellus*, Plutarch provides a detailed description of Archimedes' tomb, but historians disagree about what was actually depicted. Eratosthenes of Cyrene (*c*. 285–194 BCE) served as Librarian in Alexandria and is also renowned for his estimation of the size of the earth based on empirical measurements and calculations. He was a correspondent of Archimedes; the *Cattle Problem*, attributed to Archimedes, is addressed to him. In the *Letter to King Ptolemy*, attributed to Eratosthenes and presenting a solution to the problem of doubling the cube, there is a description of a votive monument erected to announce and celebrate the solution, presumably intended to be permanently and publicly visible. Claudius Ptolemy reported results from his own astronomical work on a votive inscription in Canopus, near Alexandria, in the tenth year of Antoninus Pius' reign (146 or 147 CE). The first words of the inscription proclaim: "To the Saviour God Claudius Ptolemy [dedicates] first principles and hypotheses of mathematics" (Jones 2005: 68–9). Serapis was presumably the god intended, the protector of Alexandria, whose Greco-Egyptian cult temple was located there. While none of these monumental testaments to mathematical achievement survive, the ancient written descriptions demonstrate that some mathematicians wished their work to be publicly represented, indicating that they thought that there was wider, non-specialist interest in and respect for mathematics.

In Book IX of his work *On Architecture*, the first century BCE Roman author Vitruvius recounts the oft-repeated anecdote about Archimedes' discovery of the mass-to-volume relationship while relaxing in a public bath (and jumping out proclaiming in Greek "*Heurēka!*" ("I found it!"). His account features elements that recur in other stories of mathematicians' work, such as the theme of a ruler requesting a solution from a mathematician. According to Vitruvius, the Syracusan ruler Hiero had asked Archimedes for help in determining whether he had been defrauded by the maker of a crown: whether the crown was pure gold or mixed with other metal (such as silver). The trope relating to a royal request and to patronage is present in other notable accounts of mathematics and seems to reinforce Euclid's message that there is "no royal path." Rather, one must consult a specialist. According to Vitruvius, Archimedes not only jumped out of the bath, but he also then ran naked into the street to make his proclamation. (This highlighting of what might be deemed to be silly behavior carries a hint of deprecation. There is a long tradition of telling slightly humiliating stories about "great thinkers." For example, in the *Theaetetus*, Plato recounts that Thales of Miletus fell into a well whilst stargazing; even a slave girl who witnessed the event thought he was a bit foolish.)

Vitruvius presents his account of Archimedes' discovery as part of a larger celebration of the contributions to mathematics made by a number of ancient Greeks. Part of his purpose, as he explains, is to highlight the work of those authors who throughout the ages had bestowed infinite benefits on humanity. Dismayed about the ways that certain activities and pursuits are valued and rewarded, while writers and those who have greatly profited others are not celebrated to the same extent as famous athletes, he proposed that intellectuals, too, should have wreaths and crowns placed on their heads, be driven triumphantly in chariots and acclaimed as heroes. He names a number of the deserving, beginning with those who made contributions to mathematics. Plato is the first, and here Vitruvius refers (without specifying the dialogue) to the geometrical solution for doubling the square discussed in the *Meno*, emphasizing that a solution cannot be achieved arithmetically.

There, Socrates quizzes a slave boy about mathematical problems. Socrates suggests that the slave boy's answers are based on recollection (*anamnêsis*); he also discusses the migration of souls. Socrates' consideration of mathematical solutions does not refer to geometrical proof, nor does he use much in the way of technical vocabulary (with the exception of the term "diagonal"). In other words, in the *Meno*, the solution does not seem to require specialist skills. Rather, the solution is provided by recollected knowledge in the mind of the slave boy: he "just knows" the answer, because he can recall knowledge from a previous existence.

That humans' mathematical knowledge is acquired by recollection or some other means aside from study is further emphasized by Vitruvius' claim that Pythagoras himself was convinced that the Muses steered him to his discovery. We are told that Pythagoras thanked the Muses and accordingly sacrificed to them. But, while Pythagoras' mathematical achievements may have been due to the inspiration of the Muses, his work was of practical benefit, for example, in the construction of buildings, including determining the proper height of steps for a staircase. In Vitruvius' view, Pythagoras' method was more helpful than that which could have been devised by craftsmen on their own. He goes on to detail how to use Pythagoras' "discovery," describing how three sticks or rods—one 3 foot long, another 4 foot, the other 5—can be put together to form a triangle with a right angle, and how a square can be formed on each straight edge, with the result that the square on the 3-foot side will be 9 feet in area, that on the 4-foot side 16, and that on the 5-foot side 25, so that the square on the 5-foot side is equal in area to the squares on the 3- and 4-foot sides. Vitruvius then goes on to explain how this principle can be used to determine the height of steps in a staircase between stories of a building; he mentions a drawing, which no longer survives, depicting how this may be done. Vitruvius, writing about architectural matters, includes a detailed discussion of the Pythagorean theorem to emphasize the practical benefit of mathematics.

Vitruvius emphasizes how human life has been improved by mathematicians, also praising Archytas and Eratosthenes for their solutions to what was understood to have been a demand made by the god Apollo to the citizens of

Delos, that the size of the altar dedicated to him be doubled. Archytas was credited as the first to have used a drawing or diagram of cylinders, and Eratosthenes as having used an instrument, the ***mesolabium***.

The representation of mathematics as being difficult contributed to the creation of legends about solutions to particular problems. Concerning the problem of doubling the cube, the author of the "Letter to King Ptolemy" (who may have been Eratosthenes) recounts the legend of King Minos desiring to double the size of his son's tomb, and the story of the Delians attempting to obey an oracle to double the size of an altar. To accomplish this, representatives were sent to Plato's Academy to get help in finding the solution. A later author, Theon of Smyrna (first part of the second century CE), reports that Plato was said to have told the Delians that the problem was given as a reproach for having neglected mathematics and geometry.

There are numerous almost hagiographical accounts in the biographies of Pythagoras placing his mathematical achievements side by side with other, almost magical or even god-like, actions. From Diogenes Laertius, we learn that he devoted a good deal of time to geometry and also discovered the musical intervals on the monochord. Pythagoras may have recognized how special was his mathematical achievement, for Diogenes Laertius reports that, according to Apollodorus—known as "the calculator," Pythagoras offered a sacrifice of oxen when he found that the square on the hypotenuse of a right-angled triangle is equal to the squares on the sides containing the right angle. Pythagoras' religious charisma is emphasized in his portrayal as a kind of prophet and the son of a god in the ancient biographies of him that survive.

In Plato's *Timaeus* we are told that the world was created through a process incorporating mathematical ideas. Later Platonism was in some instances linked to neo-Pythagoreanism; both were seen to emphasize the role of mathematics in the cosmos, and neo-Pythagoreanism regarded mathematics—including arithmetic, geometry, harmonics, and astronomy—as crucial to understanding the nature of the world. For some neo-Pythagoreans, Pythagoras acquired the status of a divine "hero." Pythagoras' fame was celebrated in various ways, including on coins depicting his portrait (Figure 7.2). The breadth of his appeal is indicated by ancient Jewish authors, including Aristobulus (second century BCE), who represented Pythagoras himself as having been influenced by Judaism. Pythagoras attracted a substantial following; Iamblichus lists the names of the seventeen "most famous" Pythagorean women, as well as 218 men.

While mathematics was a male-dominated activity in antiquity, there is evidence of women making original contributions at a very high level, as well as being involved in teaching. Music was an important branch of mathematics for the ancient Greeks, and thanks to Porphyry, in his *Commentary on Ptolemy's Harmonics*, we know something about Ptolemaïs of Cyrene (perhaps early first century CE), the only Greek female musicologist known to us at all. Her

FIGURE 7.2. Pythagoras, depicted on a coin. British Museum. Photo by Universal History Archive / Universal Images Group / Getty Images.

Pythagorean Elements of Music, with its catechistic question-and-answer format, may have been a school-teaching text.

Pandrosion, a mathematics teacher, is the dedicatee of the third book of Pappus of Alexandria's fourth-century CE *Collection*, in which he considers solutions to several problems, including the duplication of the cube, geometric construction of the harmonic mean, and the inscription of the five regular solids within a sphere. Pappus deviates from the usual practice of being complimentary in a dedication, opining that Pandrosion may not have been the most effective mathematics teacher, because some of her self-proclaimed students were incompetent. While he is critical of the efforts of some of her students, this may be due to some rivalry. No writings by Pandrosion survive, nor do other mentions of her or her work, so it is difficult to test Pappus' criticisms.

Hypatia, the daughter of Theon, was a teacher of mathematics in Alexandria. As we have seen, Synesius of Cyrene extolled her excellence. Having studied with her father and others, she taught at the Neoplatonist school of philosophy, becoming its head. She probably wrote on various topics in mathematics, astronomy, and philosophy, including planetary motions number theory and conic sections; she also edited works by other authors. Murdered by Christians, she is often represented as a pagan martyr.

Archimedes was himself depicted as a martyr to mathematics, by Plutarch, in his biography of Marcellus, the Roman military leader responsible for the siege of Syracuse, home to the renowned mathematician. Plutarch relates how Archimedes' preoccupation with mathematics led to his death at the hands of a Roman soldier; his death was regretted greatly by Marcellus.

Not all ancient accounts of mathematics focus on the work of human beings. Bees are singled out by Pappus as especially talented in producing honeycombs that have cells that are all equal, as well as similar and adjacent, and hexagonal in form. At the beginning of Book V of his *Collection*, Pappus suggests that, in their ability to produce such honeycombs, we may infer that bees "have contrived this in accordance with a certain geometrical forethought." He contrasts the style of mathematical work of bees directly with that of humans:

> Bees, then, know just this fact which is useful to them, that the hexagon is greater than the square and the triangle and will hold more honey for the same expenditure of material in constructing each. But we, claiming a greater share in wisdom than the bees, will investigate a somewhat wider problem, namely that, *of all equilateral and equiangular plane figures having an equal perimeter, that which has the great number of angles is always greater, and the greatest of them all is the circle having its perimeter equal to them.*[3]

Pappus emphasizes the generality of the mathematical problems that interest human mathematicians, in contrast to those practical problems related to honeycombs that interest bees. Indeed, before discussing the work of bees, Pappus explained that:

> Though God has given to men, most excellent Megethion, the best and most perfect understanding of wisdom and mathematics, He has allotted a partial share to some of the unreasoning creatures as well. To men, as being endowed with reason, He granted that they should do everything in the light of reason and demonstration, but to the other unreasoning creatures He gave only this gift, that each of them should, in accordance with a certain natural forethought, obtain so much as is needful for supporting life. This instinct may be observed in other species of creatures, but it is specially marked among bees.[4]

While Pappus recognized that mathematical practice was not reliant on the format of the proposition and proof, he stresses that the human contribution to mathematics was the ability to present general solutions to a mathematical problem. Note that, as we have seen in other examples, the mathematical skills of both men and bees were a divine gift.

A number of ancient Greek authors suggest that the creation and structure of the world were due to a Creator who deliberately employed mathematics in creating the world; this tradition goes back at least to Plato. Aristides Quintilianus (dates unknown, possibly late third–early fourth centuries CE) in *De Musica* (3.24) explained that:

> In the *Timaeus* the noble Plato says something like this: the craftsman who made the soul took a kind of essence intermediate between indivisible and divisible essence, and in the case of the divisible and indivisible varieties of the Same and the Other he put together [mathematical] means with the mean of those essences. When he had made a mixture of these three, he divided up the entire mixture on the basis of the following odd and even numbers: he increased the even numbers in duple ratio, up to the number 8, and the odd ones in triple ratio up to the number 27.[5]

For Plato and his followers, the role of number and mathematics in the creation of the world was crucial.

Other ancient authors, not themselves known to have been mathematicians, also give a sense that those engaged in mathematical pursuits were somewhat different from others, and not always in an enviable way. So, for example, in *Clouds*, Aristophanes jokes about mathematicians engaged in astronomy and geometry. In *On Listening to Lectures*, Plutarch pokes a bit of fun at those who display their mathematical expertise, complaining that those young men who boastfully use words like "diagonal," seeking to impress others, would be better off thinking about how to live a wholesome life. Yet, lest we think that Plutarch had something against mathematicians, he engages very closely with mathematical problems and solutions elsewhere, including in his *Commentary on the Generation of the Soul in the* Timaeus, where once again mathematics is represented as essential to the nature of the universe.

MATHEMATICS AS A SPECIFIC SORT OF SKILL/ ACTIVITY, NOT ALWAYS RESTRICTED TO SPECIALISTS

While mathematics was implicated by some in the creation of the world itself, there were also more light-hearted uses of mathematical skills and knowledge. A number of poems, including the so-called *Cattle Problem*, the epigrammatic problems in the *Greek Anthology*, and the isopsephic (equal pebbles, used for counting; the Greek word for "pebble" is *psephos*) poems of Leonides of Alexandria (first century CE) indicate that engaging in mathematical problem-solving was sometimes a recreational and entertaining activity.

An ancient question-and-answer game or contest featuring numbers and counting is depicted as having taken place between the two oldest known Greek poets, Homer and Hesiod. Among the questions the interlocutor named "Hesiod" asks is the following: How many Achaeans went to Troy together with the sons of Atreus? The character named "Homer" answered with a calculating-problem: there were fifty hearths, and in each there were fifty spits, and on each fifty pieces of meat, and three (times) three hundred Achaeans were around each piece of meat. Sadly, the original text does not survive, but there is an attempt—possibly by a later annotator—at a calculation. That this mathematical problem is included in a story about who is the best of the two famous poets gives a strong signal that mathematics was well embedded within broader Greek culture.

Just as it seems that it was reasonable to depict famous poets as mathematicians, famous mathematicians could also be accomplished poets. An important example of mathematics presented as poetry is found in the so-called *Cattle Problem* attributed to Archimedes; whether or not he was responsible for it is unknown. The problem is noteworthy for the difficulty of the mathematics involved, worthy of the great mathematician. A seemingly simple and practical problem about counting herds of different colored cattle is set out in meter; the problem requires that eight unknown quantities are found. (The *Cattle Problem* was solved satisfactorily only in the twentieth century; the solution required the use of computers. We do not know whether the ancient author of the poem thought it was solvable.)

The *Cattle Problem* is presented as a poem. The mathematical problem is itself tightly woven into the poetic format. The text opens this way: "a Problem [*problēma*] which Archimedes devised in epigrams [*epigrammata*], and which he communicated to students of such matters at Alexandria in a letter to Eratosthenes of Cyrene"; we know from other sources that Archimedes, living in Syracuse, did in fact correspond with Eratosthenes. Here, the poetic, epigrammatic, format of the problem is emphasized; no apology or explanation is given for the genre. Furthermore, no ancient prose-setting of the mathematical contents of the *Cattle Problem* is known. Both the author of the *Cattle Problem* and the intended recipients are portrayed as people who will be accustomed to dealing with poetic texts.

There are other mathematical texts which assume specialized mathematical language and formats (treated elsewhere in this volume), and there were mathematical texts which appear to have been devised, at least partly, to be entertaining, especially as many of them are riddles presented in short epigrammatic poems. The *Greek Anthology* includes poems ranging from the Classical to the Byzantine periods, and more than forty mathematical problems presented as epigrams. Many of these problem-poems are thought to have been collected by Metrodorus of Tralles (usually referred to as a grammarian, *c.* 550–600 CE, although his dates are not certain). Historians have suggested that the mathematical problems likely were devised (and even the poems possibly composed) much earlier. This raises the question of why a grammarian would be interested in mathematical material. There are other poems in the *Anthology* that involve counting. Certainly, some of the texts that we would identify as "mathematical" were written by people who were probably not specialists in mathematics.

These mathematical epigrams pose problems to be solved. Typically, no solution was offered, leaving the reader to puzzle it out for themselves. In order for the reader or auditor fully to comprehend the mathematical epigrams, they must engage in mathematics, they must "do" math. (Whether or not readers would have been expected to consult other texts or people—including teachers—to arrive at a solution is not clear.) These arithmetical problem-poems represent the use of mathematical skills as something potentially enjoyable, entertaining and rewarding.

Many of the mathematical problems in the *Greek Anthology* appear at first reading to be simple and mundane, requiring the counting of pieces of fruit or bowls, yet several are mathematically sophisticated. Even a brief consideration of them shows that many of these mathematical epigrams had correspondences with well-known myths and legends; several have clear intertextual references to other poetry, including the Homeric and Hesiodic poems, as well as allusions to myth more generally. Given the abundant cultural allusions and references to myths, specific signals to other texts were probably not necessary, especially when depictions of myths were readily accessible in vase paintings and other art forms.

There are some epigrams in the *Anthology* that, although not being mathematical problems, encourage readers to count. Leonides (sometimes spelled Leonidas) of Alexandria, reportedly an astrologer or astronomer during the reign of Nero (54–68 CE), composed a number of isopsephic poems. More than forty poems are credited to him, several having been offered as birthday gifts to members of the emperor's family, indicating patronage links. The letters of the Greek alphabet had numerical value ($\alpha = 1$, $\beta = 2$, etc.), and therefore every letter in a word or line of verse could be assigned a value as well.

Leonides' isopsephic poems invite readers to convert the letters and words of each verse into numbers, which are then counted and compared, to discover that the verses are equal numerically. The numbers that result are typically large (in the thousands), and it is easy to imagine that the challenge of counting so high also offered some special pleasure, something like talking about a "gazillion." Archimedes' *Sand-Reckoner* provides evidence of mathematicians' fascination with large numbers, and while Leonides' epigrams do not require a great deal of mathematical skill or experience, they do rely on the reader's active engagement, and presume a degree of knowledge about numbers and counting. And, as is typical for epigrammatic poems, there is a degree of intellectual challenge and playful surprise.

Book XIV of the *Greek Anthology* is devoted to riddles, oracles, and mathematical problems. The mathematical poems are a form of "story-problem" requiring some level of comfort with mathematics and also expecting a certain degree of familiarity with poetic traditions. Like the *Cattle Problem*, those in the *Greek Anthology* were presented without solutions; the epigrammatic form seems to have been particularly suitable for posing problems. But unlike the *Cattle Problem*, those in the *Greek Anthology* were apparently thought to be solvable.

Judging from their content, many of the mathematical problems may have originated as early as the fifth century BCE or even earlier. Plato, in the *Laws*, presented the Athenian stranger, the Cretan Clinias, and the Spartan Megillus discussing, amongst other topics, questions having to do with the proper education of children. The importance of learning arithmetic, counting, and calculation—and of having fun while doing so—is stressed, when the Athenian states that:

> First, there should be calculations of a simple type devised for boys, which they should learn with amusement and pleasure, such as the distributions of apples and crowns wherein the same numbers are divided among more or fewer, or distributions of the competitors in boxing and wrestling matches. . . . the boys should play with bowls containing gold, bronze, silver and the like mixed together, or the bowls may be distributed as wholes. For, as I was saying, to incorporate in the pupils' play the elementary applications of arithmetic will be of advantage to them later in the disposition of armies, in marches and in campaigns, as well as in household management, and will make them altogether more useful to themselves and more awake.[6]

The passage goes on to emphasize the importance of metrology and of geometry, and the practical importance of a mathematical education is emphasized. A number of the mathematical poems collected in the *Greek Anthology* fit the requirement set out in the *Laws*, and deal with problems related to apples, ornamental headgear, and bowls, as well as other objects made of different metals. The emphasis on calculation stresses the sort of practical problem that might be encountered in everyday life, even when couched in the framework of familiar and traditional myth, and epic poetry.

As we have seen before, mathematics is of special use to royalty. Herodotus provided a long list of all the items that Croesus, king of Lydia from *c.* 560 to 546 BCE, until his defeat by the Persians, had sent to the oracle in Delphi; many have the exact weight specified. Epigram 12 in Book XIV of the *Greek Anthology* specifies that he dedicated six bowls weighing six minae, each one drachm heavier than the other. Here, a legendary account of Croesus is given specific mathematical meaning through reference to the practical problems of weighing different bowls, a problem very similar to those mentioned by Plato in the *Laws*, where calculations about bowls made of the same or of different metals serve as an educational tool. Epigram 49 also refers to valuable materials fashioned as royal headgear:

> Make me a crown weighing sixty minae, mixing gold and brass, and with them tin and much-wrought iron. Let the gold and bronze together form two-thirds, the gold and tin together three-fourths, and the gold and iron three-fifths. Tell me how much gold you must put in, how much brass, how much tin, and how much iron, so as to make the whole crown weigh sixty minae.[7]

This problem, involving the solution of several unknowns, also has mathematical resonances with Archimedes' *Cattle Problem*, but on a different scale. And there is also an allusion to myth: the crown may well have belonged to Croesus or some other king.

The mathematical problems in the *Greek Anthology*—couched in terms of traditional myths, dealing with wealth in the form of bowls made of precious metals or even crowns—also refer to potentially real objects and seemingly practical problems. While it is difficult to imagine real-life applications of some of the complicated problems presented in the poems, the counting of objects like bowls and apples was seen—even by Plato—as something necessary to learn for everyday life. Determining the weight of precious metals was crucial to ascertain their value, as we know from the story of Archimedes shouting, "*Heurēka!*"

The mathematical riddle collection in the *Greek Anthology* is evidence that mathematics was not always treated and taught as a formal science only by specialists. It is possible to imagine learned enthusiasts—interested both in mathematics and in poetry—having fun with these epigrams, engaging in a sort of mathematical recreational activity. Indeed, the composition, hearing, or reading, and solving of these required a number of different skills: the solving of the problem would have been one (mathematical) challenge; the setting-out of the problem in a poetic epigram (with allusions to other texts, including poetry, myth, and mathematics), and the informed multi-layered "reading" of these epigrams would have presented others. One can well imagine that this is just the sort of "party piece" that would have won favor among the educated men who attended the dinner parties known as symposia. Riddling poems of various kinds were a standard form of sympotic literature, whether actually delivered or composed and collected as the sort of thing that might be offered. These poems suggest that solving mathematical problems was an enjoyable and entertaining activity to be shared with others.

VISUAL REPRESENTATIONS AND TANGIBLE OBJECTS

Various visual representations of activities involving mathematics survive from several ancient cultures, produced using a variety of media, such as paintings, coins, mosaics and stone reliefs. A number of references in ancient works refer to objects and monuments that were created to represent mathematical achievements—like Archimedes' tomb—but these no longer survive, if they ever did physically exist. Many of the extant images depict men doing things with an object or tool used for counting, measuring or calculating. These are likely to be representations of practical numeracy, and probably not of higher mathematics. Whether this is the case simply because it may be easier to represent tangible objects rather than abstract ideas is an open question.

Among the inscribed clay tablets from Old Babylonia (*ca.* 1800–1600 BCE in what is now Iraq) in the Yale Babylonian Collection are several concerned with mathematics (Figure 7.3). Trainee scribes did some of their work on (approximately) round or square clay tablets, with diameters ranging from 8 to 12 cm. Known as "hand tablets," they fit comfortably within the palm, and were erasable, thereby suitable for the practice of handwriting and mathematics. The writing is bigger and sloppier on some than on others; presumably, scribes' writing shrunk and became tidier over time, through practice. Other inscribed tablets also depict diagrams, such as a diagram of a circle and the computation of its area, and that of a trapezoid and the calculation of area.

FIGURE 7.3. Clay tablet with a drawing of a square with a diagonal and inscribed numbers. Yale Peabody Museum of Natural History Catalogue Number YPM BC 021354. Photo by DeAgostini / Getty Images.

Even though descriptions of diagrams played a crucial role in ancient Greek mathematics, few examples of drawn diagrams survive from Greek antiquity. There are numerous detailed descriptions of lettered diagrams, but the ancient manuscripts in which they would have originally appeared no longer survive (see elsewhere in this volume).

Various kinds of mathematical work and tasks utilizing mathematical skills required the use of tools or instruments. A number of verbal accounts and representations of mathematicians and mathematical practice draw attention to things invented or used by practitioners. Some surviving visual representations of mathematics and of those engaged in tasks requiring mathematical ability depict objects associated with such work, reminding us that not all mathematics was abstract.

ROPE MEASURING AND RECKONING

Land surveying and measurement were done using rods, poles, and cords of rope knotted at regular intervals. Clement of Alexandria (born *c.* 150; died between 211 and 216 CE) reports that Democritus referred to the ability of Egyptian *harpedonaptae* to compose lines by drawing. *Harpedonaptae* may mean cord- or rope-stretchers but it is a rarely occurring Greek term, and its use is not entirely clear. We do know that knotted ropes were used in different contexts for different sorts of measurements, but we do not really know what these *harpedonaptae* were supposed to have done. There is evidence in Eighteenth Dynasty (1550/49–1292 BCE) art depicting the use of cords, possibly for land measurement and surveying. Images of Egyptian "rope-stretchers" are found on a number of wall paintings in Theban tombs, including those of Menna, Amenhotepsesi, Khaemhat and Djeserkaraseneb (see Chapter 2). However, the interpretation of these scenes is not straightforward, and some scholars suggest that ropes were used to orient rather than measure. Some Egyptian hieroglyphs may refer to activities such as counting or measuring, such as the sign that may represent a measuring rope (Rossi 2007).

The ancient Greek historian Herodotus reports how the Persian king Darius explained the use of a knotted rope as a sort of calendar, when he instructed the Ionians how to determine the passage of time: "Reckon from the day you see me marching away and untie one knot each day. If all the days represented by the knots have passed and I am not back, head for home."

PEBBLES AND ABACI

Familiarity with counting and calculating practices from as early as the sixth century BCE are attributed by much later sources to the Athenian statesman Solon, who is supposed to have compared tyrants' men to the pebbles of an abacus. The etymology of "calculation" can be traced to Greek and Latin words for pebbles.[1] And, as discussed in Chapter 1, evidence for the use of pebbles in counting and calculating survives in objects, as well as in written texts. Pebble-counting for calculations too difficult to be done in the head (or on the fingers) is mentioned in the fourth century BCE by the Athenian statesman and orator Demosthenes (385/4–322 BCE).

The "Darius" krater (340–20 BCE), named after the Persian king depicted in a central portion of the vase, illustrates a man dressed in the Greek fashion seated at a table (Figure 7.4). He appears to be using pebbles to count or calculate and holds a wax tablet, on which is written TAΛATA:H presumably meaning *tal'anta hekaton*, "a hundred talents," referring to the Greek monetary unit, the *talent*, with the letter *H* in acrophonic notation designating the first letter of the word for one hundred (Smith 1925: vol. 2, 161–2; Cuomo 2001: 10–13). This may record the result of a transaction, or what was expected. Another man stands opposite, wearing distinctly non-Greek headgear and what appear to be trousers, in the Persian fashion, with outstretched arms proffering a bundle. Other figures, empty-handed, are also lined up with their arms out, perhaps beseeching. The scene has been interpreted as depicting the collection

FIGURE 7.4. Detail from the "Darius" krater. Museo Archeologico Nazionale, Naples, number 81947; H3253. Photo by Wikimedia Commons.

of taxes or tribute from Persians to a Greek. (This portion of the painting is directly below an inscription describing "Persians," *Persai*.)

The table itself appears to function as a sort of reckoning board or abacus, on which are marked Greek letters representing numerals, to aid in calculating large numbers. White pebbles or counters below the letters may indicate the number of each quantity unit, but interpreters are not univocal in their readings. The vivid portrayal of the Greek figure actively tallying tribute is ambiguous when set against the backdrop of the historian Herodotus' portrayal of Darius as a shopkeeper, preoccupied with the handling of money; nevertheless, both tasks required sufficient numeracy and were aided by counting objects.

Herodotus tells us that Greeks wrote and calculated from left to right while Egyptians did so in the other direction. He describes the Greek method of counting with pebbles in what were presumably vertical columns, and so likely a form of abacus. Two references in Aristophanes' *Wasps* refer to the physical act of counting and may suggest the use of an abacus (one referring to a stone, or *lithos*, another referring to *psephoi*, pebbles), and the use of a stone slab with pebbles to count votes in court is described in Aristotle's *Constitution of the Athenians*.

Some ancient abaci survive, including the bronze Roman "pocket" abacus held in the Cabinet des Médailles of the Bibliothèque nationale in Paris. We also have visual representations of people using what appears to be an abacus, including an Etruscan cameo showing a seated man holding what may be an abacus, also in the Cabinet des Médailles, and the Testamentum relief at the Museo Capitolino in Rome, depicting what has been interpreted as a slave with an abacus (Figure 7.5; D'Ambra 1995).

The practical value of the abacus—and the mathematical practices it supported—may be a reason why we have a variety of evidence. There must have been many more examples of pebbles used for counting and calculating as well as abaci in addition to those known to survive.

SUNDIALS

Several mosaics (primarily Roman) depict individuals or groups of people with objects, such as sundials. One of these, a third-century CE Roman mosaic (Figure 7.6), shows a sundial held by a man, possibly intended to be Anaximander of Miletus (died *c.* 547 BCE). Anaximander was credited by Diogenes Laertius as having been the first to invent or discover the *gnomon*, a shadow caster, which he set up in Sparta, to mark the solstices and equinoxes. How this mosaic came to be associated with Anaximander is unclear; it may be due to knowledge of the story of his invention of the dial.

The construction of sundials required mathematical knowledge. There are over six hundred ancient sundials that survive, and we have excellent examples of the types described by Vitruvius in Book IX of his *De architectura*.

FIGURE 7.5. Testamentum relief, early 2nd century CE. Museo Capitolino. Photo © Ann Raia, 2005. From the VRoma.org image archive.

Another Roman mosaic (found at Pompeii in 1897), is thought to depict Plato's Academy (Figure 7.7). The image, sometimes referred to as the "mosaic of the seven sages" or philosophers, shows seven bearded men grouped around what may be a celestial globe resting in a box-like stand—possibly related to the type described by Ptolemy in the *Syntaxis*, where he gives instructions for constructing a star-globe. Above it and resting on a column is a sundial, probably of the spherical or the conical type. Various proposals have been made regarding identification of the people portrayed; that they are depicted near objects associated with mathematics (especially astronomy), has led to an assumption that at least some of them were engaged in mathematical practices. A group of people—rather than a lone individual—is shown, also suggesting or even emphasizing that such mathematical endeavors involving the study of the heavens rely on the work of more than a single individual. The group nature of the project is made clear, while symbols of mathematical knowledge are displayed. In fact, many (if not most) of the images depicting mathematical objects being used portray some sort of group situation, even when the group is comprised of only two people. In some cases, for example the Testamentum relief, we do not really know what the interaction involved; this is left to the

FIGURE 7.6. Roman mosaic depicting a man holding a sundial. Landesmuseum Trier, Germany. Photo by PHAS/Universal Images Group / Getty Images.

imagination, or to our presumed knowledge of the relationships of the individuals depicted, and their role in the household.

The mosaics—with representations of the globe and sundials—simultaneously illustrate both the more theoretical (the idea that the heavens are spherical in shape) and the more practical aspects (for example, time-finding skills) of ancient mathematics, and signal that mathematics itself, in antiquity, represented a broad range of types of knowledge and skills. Herodotus' assertion that the

FIGURE 7.7. "Seven Sages" or "Plato's academy" mosaic from the Villa of Titus Siminius Stephanus, Pompeii, first century BCE–first century CE. Museo Archeologico Nazionale, Naples, Inv. No. 124545. Photo by Leemage / Corbis / Getty Images.

Greeks had learned about the *polos* (sundial) and the *gnomon* and the twelve parts of the day from Babylon, rather than receiving it from a god or inventing it for themselves, underscores the willingness of some ancient historians to represent their debt to other cultures, and the sharing of mathematical knowledge across geographical and linguistic boundaries. Taken together, ancient textual and visual representations of mathematics provide us with a richer picture of the place of mathematics in ancient culture than we would have if we relied on only one type of evidence.

This short survey of visual representations of mathematics has tended—perhaps inevitably—to focus on practical applications of mathematical knowledge and skills, while textual material represents other aspects and characteristics, not only the abstract nature of some types of mathematics, but also many of the wider cultural associations. Ancient visual and textual representations of mathematics underscore the point that while there was no royal road to mathematical prowess, appreciation of mathematical skills was not confined to specialists.

NOTES

Introduction

1. For more insight into the question of reader, look into books such as those by Kuper and Kroeber and Kluckhohn, mentioned above, as well as classics such as Edward B. Tylor, *Primitive Culture* ([1871] 1929, 2 vols), where the word "culture" was introduced in its modern anthropological sense and Ruth Benedict's *Patterns of Culture* ([1934] 1959), in which the idea of "cultural relativism" was developed. These works, imperfect though they may be by modern standards, are still fascinating and insightful.
2. Proclus, *In primum Euclidis Elementorum librum commentarii*, Friedlein, p. 46, translation by Glenn R. Morrow, *Proclus: A Commentary on the First Book of Euclid's* Elements (Princeton, NJ: Princeton University Press, 1970), 38. Unless otherwise noted, all translations from Proclus will be taken from Morrow's edition.
3. For details on the liberal arts, consult David L. Wagner, *The Seven Liberal Arts in the Middle Ages* (Bloomington, IN: Indiana University Press, 1986); for more on the history of education in ancient times and the Middle Ages, see H. I. Marrou, *A History of Education in Antiquity*, George Lamb (trans.), (Madison, WI: University of Wisconsin Press, 1982).
4. This is Aristotle's categorization, as Ptolemy points out.
5. This whole issue of anachronism in the history of mathematics is the subject of a recent book edited by the eminent Newton scholar, Niccolò Guicciardini.
6. The Vatican Euclid, Vat. Gr. 190, called P; a digital copy of the manuscript may be found here: https://digi.vatlib.it/view/MSS_Vat.gr.190.pt.1 (accessed March 9, 2020).
7. This diagram appears in the *Papyrus Oxyrhynchus* 29 and is one of the oldest extant diagrams from Euclid. For more details about this papyrus, see the website: http://www.math.ubc.ca/~cass/Euclid/papyrus/ (accessed March 12, 2020).
8. For this substantial material about China, I owe thanks to Joe Dauben, and for the material about India, to Kim Plofker.

9. The passage is in the 1637 edition, pp. 305–6:

    [p]Où je vous prie de remarquer en passant que le scrupule que faisaient les anciens d'user des termes de l'arithmétique en la géométrie, qui ne pouvait procéder que de ce qu'ils ne voyaient pas assez clairement leur rapport, causait beaucoup d'obscurité et d'embarras en la façon dont ils s'expliquaient.

    Here I beg you to observe in passing that the scruple the ancients had about using terms from arithmetic in geometry—which could only proceed from the fact that they did not see clearly enough the relationship between them—caused much obscurity and embarrassment in the manner of their explanations. (My translation)

Chapter One

1. Studying the investigation of numeracy in the ancient world had been taken out academically only in the past several years. Central texts so far have been published by Melissa A. Bailey, Seraphina Cuomo, Raviel Netz, and Eleanor Robson. This chapter owes much to their texts, even where they are not explicitly cited.
2. RRC = M. H. Crawford, *Roman Republican Coinage* (London: Cambridge University Press, 1974).
3. For example, in third millennium BCE Mesopotamia, scribes used the same lists of division to quantify grain, to calculate theoretical laboring days, and to measure land (Robson 2008a: 67–84).
4. DK = H. Diels and W. Kranz eds., *Die Fragmente der Vorsokratiker*, 6th ed. (Berlin: Wiedemann, 1952). For English translation, see Seaford (2004), 269.
5. For example, *tetradrachma* is literally "four *drachmas*"; *triobol* is "three *obols*," also called *hemidrachma*, "half a *drachma*," since the *drachma* was divided into six *obols*; *tetartēmorion* is a quarter of an *obol*, *tetarton* meaning "a fourth" and *morion* "a part."

Chapter Two

1. From now on, all the dates are given BCE, except when specified.
2. Precise numbers are given by the database Trismegistos, available online: https://www.trismegistos.org/index.php (accessed September 20, 2019). At this date, the database inventory gives 2,027 Hieratic papyri and 4,792 Demotic papyri, 9,307 Demotic ostraca and 4,305 Hieratic ostraca (but this number overlaps the previous one because many of these are in Demotic with a few words or sentences in Hieratic). Note that Hieroglyphic papyri are rare; they are mainly copies of texts engraved or painted on the walls.
3. Graves provide perfect climatic conditions for the preservation of papyri. The important towns were located along the Nile where the atmosphere is more humid, and modern cities are built and excavations are hardly possible.
4. The *Late Egyptian Miscellanies*, according to the most recent interpretations, are not school exercises, see Imhausen (2016), 143.

Chapter Three

1. The presentation of cuneiform sources is based on the works of many authors, among them are Jens Høyrup, Jöran Friberg, Eleanor Robson, Robert Middeke-Conlin, Cécile Michel, and Mathieu Ossendrijver. Only a short selection of main works are provided in this bibliography.
2. Powell (1987–90).
3. VAT 12593, MS 3047, Feliu 2013, A 681, CUNES 50-8-1. See information on these tablets (date, provenience, publication, possible photo, and copy) on the CDLI website, https://cdli.ucla.edu/, respectively under the following numbers: P010678, P252059, P464229, P222256, P274845. The same holds for the other tablets quoted in this chapter. The reader is invited to refer to the CDLI website under the corresponding CDLI number.
4. Metrology: 1 ninda ←12— kuš ←3— šu-du-a ←10— šu-si, with 1 ninda ≈ 6 m (see cdli:wiki, available online: http://cdli.ox.ac.uk/wiki/doku.php?id=ed_iii_metrological_systems, accessed October 21, 2021).
5. CDLI numbers: P250375, P218053, P213163, P213165, P020479.
6. The calculations can be checked by the reader using the online Babylonian Calculator *Mesocalc* developed by Baptiste Mélès, available online: http://baptiste.meles.free.fr/site/mesocalc.html (accessed October 21, 2021).
7. CDLI number P253067.
8. In modern terms: noting *n* a number and *n'* its reciprocal, i.e. n =1/n, the triple {(*n*-*n'*)/2, 1, (*n*+*n'*)/2} is Pythagorean. Indeed, the sum of the squares of the two formers is equal to the square of the later.
9. The symbol "//" represents the cuneiform sign 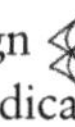 used in Hellenistic mathematical texts both for separating numbers, and for indicating empty sexagesimal positions.

Chapter Four

1. Perfect in the usual sense. As a technical term, also owing to the Pythagoreans, "perfect" means something else. A perfect number (*teleios arithmos*, in Greek), for them, was one which is equal to the sum of its proper divisors. Thus, for example, 28=1+2+4+7+14 (in this sense, 10 is not perfect!). These numbers were investigated by Euclid and Euler and are still of interest in modern number theory, mostly because, to this day, no one has proven that there are more than a finite number of such numbers or that there is even one which is odd.

Chapter Five

1. A *bēru*, in this case, is a unit of distance. As we saw earlier, is also a unit of time. In fact, 1 *bēru* of time is the time it takes to go a distance of 1 *bēru*!

Chapter Six

1. In one instance (K.2520+ rev. 43), a Neo-Assyrian glass recipe identifies itself as an *epištu*, which may also be translated as "procedure."
2. For a recent edition and discussion of HS 245 (formerly HS 229) and related texts, see Horowitz (2014), 214ff.
3. Available online: http://build-oracc.museum.upenn.edu/glass/akk-x-stdbab?xis=akk.r0000cb (accessed February 26, 2023).
4. It is possible that the pronoun was seen as redundant, as the second person subject is already made explicit in the suffix */-ka/* affixed to the infinitive form of the verb *epēšu*.
5. MUL.APIN II i 68, for which see Hunger and Pingree (1989), 87.
6. Finkel (2000), 146–7, points out that: "It is a well-known characteristic of Mesopotamian medical literature that recipes do not as a rule specify quantities; listing drug names usually will suffice for a professional practitioner who will have known through experience or instinct how much was needed in a particular case."
7. For a digital edition of the text, see: http://oracc.museum.upenn.edu/dccmt/P254608 (accessed February 26, 2023).
8. See Robson (2007a), 138, for a full translation of BM 96957+VAT 6598.
9. An overview of technological changes in the late second millennium BCE may be found in Robson (2001a).
10. For a recent survey of the horse in Mesopotamian textual sources, see Marzahn (2019).
11. It should be noted that the Kikkuli text is one of a number of Hittite horse-training texts, for a discussion of all horse-training manuals and their textual transmission, see Raulwing (2005), 61–2.
12. For a discussion of the history of the Kikkuli text's textual transmission, as well as the various Indo-Aryan terms discussed here, see Raulwing (2005).
13. As Kammenhuber (1988), 44 states: "The Mittani-Hurrians did no more understand why there were (Indo-)Aryan round numbers only for odd numbers. So, they added new terms for 2, 4, 6, in Hurrian and adapted to the new situation."
14. Buccellati (2008), 158; note that I have modified "three laps" in Buccellati's translation to "three rounds" in order to match the standard translation for *t(e) riwartanna*. Additionally, I have added quotation marks for "mile(s)" in order to indicate that these miles represent the unit DANNA rather than a mile as we understand it today.
15. The fifteen-meter IKU was suggested in Melchert (1980).
16. Note that DANNA and IKU are logograms which, in the case of the Hittite instructions, would have been understood as nouns. The standard Akkadian "mile," or *bēru* (written DANNA logographically), is between 10.69 and 10.8 km, and was the assumed distance employed in the Kikkuli text in Kammenhuber (1961). For a discussion of these distance discrepancies, see Raulwing (2005), 64–5.
17. Translation by author from VAT 9542 rev. 2–4.
18. Equally, the Assyrian texts do not preserve Indo-Aryan terms with the exception of the Akkadian term *susānu*, "horse-trainer" derived from LÚ*assussanni*.
19. VAT 9751 rev. 9–15, translation by author.

20. Among these, the stone list ***abnu šikinišu***, which explores the ritual and medicinal properties of materials classified as "stones"; see Schuster-Brandis (2008).
21. See Escobar (2018) for an essay on the historiography of the Neo-Assyrian glassmaking recipes.
22. A full discussion of ***bēru*** is offered below within the context of glassmaking procedures; see Steele (2019b) for a recent survey of short time in Mesopotamia.
23. For a digital edition of K.4266+ by the author, see: http://oracc.org/glass/P395468 (accessed February 26, 2023).
24. The identification of one kisal to 9.6 grams is briefly discussed in Heeßel (2018), 330, following Frahm (2009).
25. The term ***tērsītu***—here an intermediary vitreous material used for making a variety of colored glass—is also attested in astronomical contexts from Late Babylonian sources, where the term is translated as "computed tables."
26. See, for example, BM 29371, discussed in Brown, Fermor, and Walker (1999–2000), 144ff.
27. See Oppenheim et. al. (1970), 33, for the standard translation, in which ***apu*** is translated as "leather strap." For a complete digital edition of only text K.203+, see: http://oracc.org/glass/P393786 (accessed February 26, 2023).
28. AO 6555, 16–24; translation Robson (2007a), 177–9; Akkadian metrological terminology added in parentheses.

Chapter Seven

1. I dedicate this chapter to the memory of Andrew Barker, Wilbur Knorr, and Ian Mueller. I thank Alberto Bardi, Arthur Harris, Charlie Pemberton, Hanneke Salisbury, and the volume editor, Michael N. Fried, for their comments on earlier drafts, and Charlie Pemberton and Toby Bryant for their help with sourcing images.
2. Morrow ([1970] 1992), 24, who was unable to find the relevant passage in Aristotle, suggested that Proclus might have had in mind Plato's *Republic* 528b–c.
3. Trans. Thomas (1941), vol. 2, 593, his emphasis.
4. Ibid., 589.
5. Trans. Barker (1989), 526.
6. Trans. Thomas (1939), vol. 1, 21–3.
7. Trans. Drabkin, in Cohen and Drabkin (1948), 26. The solution given by W. R. Paton (in the Loeb edition of the *Greek Anthology*, 1918), vol. 5, 51, is: gold 30½, brass 9½, tin 14½, iron 5½; see also Taub (2017), 46.

APPENDIX

CHRISTINE PROUST

For most readers, the world of the ancient Near East is far less familiar than the Classical world or even Egypt. This is true both about the general history and geography of Mesopotamia and about the scholarship connected with its mathematical history. Thus, we thought the general information contained in this appendix might help the reader follow some of the references made in those chapters focused on Mesopotamia.

1. DATES

Periods covered by this section (estimated dates before 1000 BCE are that adopted by the CDLI according to the middle chronology; Table A1)

TABLE A1. Periods in the history of Mesopotamia

Periods	*Dates* BCE
Early Dynastic I–II	*c.* 2900–2700
Early Dynastic IIIa (Fara)	*c.* 2600–2500
Early Dynastic IIIb	*c.* 2500–2340
Old Akkadian (Sargon Dynasty)	*c.* 2340–2200
Ur III (Third Dynasty of Ur)	*c.* 2100–2000
Early Old Babylonian (Rulers of Isin-Larsa)	*c.* 2000–1900
Old Babylonian (Hammurabi Dynasty)	*c.* 1900–1600
Persian (Achaemenid rulers)	547–331
Hellenistic (Seleucid rulers)	323–63

2. MAP

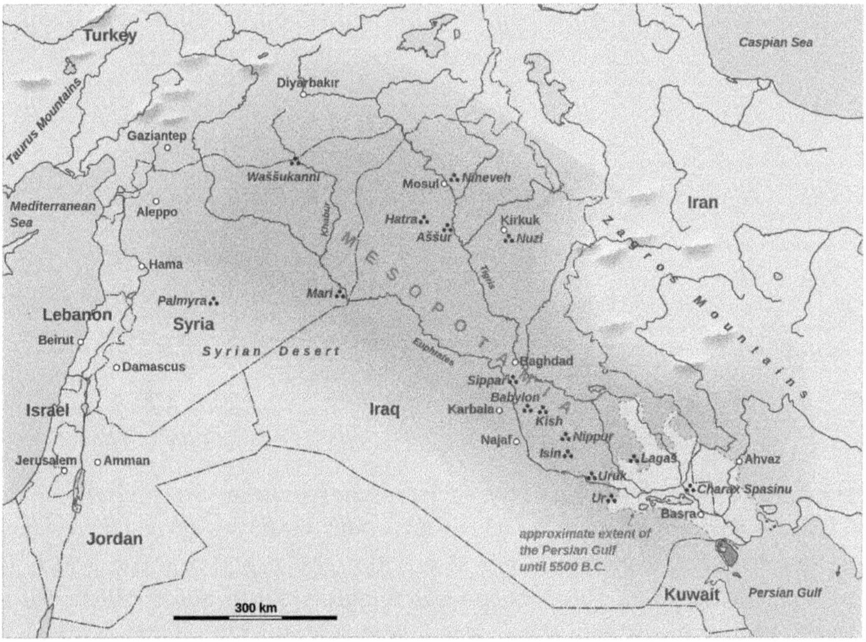

FIGURE A1. Map of Mesopotamia. Courtesy of WikiCommons. Public Domain.

COMMON ABBREVIATIONS

CDLI Cuneiform Digital Library Initiative (https://cdli.ucla.edu/)
ED Early Dynastic
LB Late Babylonian (Persian and Hellenistic periods)
OAkk Old Akkadian (Sargon dynasty)
OB Old Babylonian
SPVN Sexagesimal Place-Value Notation

BIBLIOGRAPHY

Alföldi-Rosenbaum, E. (1971), "The Finger Calculus in Antiquity and in the Middle Ages: Studies on Roman Game Counters I," *Frühmittelalterliche Studien*, 5 (1): 1–9.

Allen, James P. (2002), *The Heqanakght Papyri*, New Haven, CT: Yale University Press.

Allen, James P. (2013), *Egyptian Art in the Age of the Pyramids*, New York: Metropolitan Museum of Art.

Al-Rawi, Faruk N. H., and Andrew R. George (1991), "Enuma Anu Enlil XIV and other Early Astronomical Tables," *Archiv für Orientforschung*, 38/9: 52–73.

Amrhein, Annastasia, Clare Fitzgerald, and Elizabeth Knott, eds. (2019), *A Wonder to Behold: Craftsmanship and the Creation of Babylon's Ishtar Gate*, Princeton, NJ: Princeton University Press.

Andreau, Jean, Pierre Briant, and Raymond Descat, eds. (1997), *Economie antique: Prix et formation des prix dans les économies antiques*, Toulouse: Saint-Bertrand-de-Comminges.

Artmann, Benno (1999), *Euclid: The Creation of Mathematics*, New York: Springer.

Aruz, Joan, Kim Benzel, and Jean M. Evans, eds. (2008), *Beyond Babylon: Art, Trade, and Diplomacy in the Second Millennium B.C.* New Haven, CT: Yale University Press.

Association pour le développement de l'archéologie en Languedoc-Rousillon (ADAL), ed. (2005) *Les Équidés dans le monde méditerranéen antique. Actes du colloque organisé par l'École française d'Athènes, le Centre Camille Jullian, et l'UMR 5140 du CNRS, Athènes, 26–28 Novembre 2003*, Lattes: Édition de l'Association pour le développement de l'archéologie en Languedoc-Roussillon.

Asper, Markus, ed. (2013), *Writing Science: Medical and Mathematical Authorship in Ancient Greece*, Berlin: De Gruyter.

Attinger, Pascal (2017), "Edubba'a (5.1.1.)," *Cuneiform Digital Library Initiative*. Available online: https://cdli.ucla.edu/dl/pdf/P464238.pdf (accessed September 20, 2019).

Aujac, Germaine (1987), "The Growth of an Empirical Cartography in Hellenistic Greece," in J. B. Harley and David Woodward eds., *The History of Cartography,*

vol. 1: Cartography in Prehistoric, Ancient, and Medieval Europe and the Mediterranean, 155–7, Chicago, IL: University of Chicago Press.
Bagnall, Roger S., Kai Brodersen, Craige B. Champion, Andrew Erskine, and Sabine R. Huebner, eds. (2013), *The Encyclopedia of Ancient History*, Malden, MA: Wiley-Blackwell.
Bailey, Melissa A. (2012), "Calculative Objects: Sustaining Symbolic Systems in the Ancient Mediterranean," in John K. Papadopoulos and Gary Urton eds., *The Construction of Value in the Ancient World*, 516–35, Los Angeles, CA: University of California Press.
Bailey, Melissa A. (2013), "Roman Money and Numerical Practice," *Revue Belge de Philologie et d'Histoire*, 91 (1): 153–87.
Barbotin, Christophe (2008), *Âhmosis et le début de la XVIIIe dynastie*, Paris: Éditions Pygmalion.
Bardi, Alberto (2019), "Mathematics and Cultures across the Chessboard," in Bharath Sriraman ed., *Handbook of the Mathematics of the Arts and Sciences*, 2421–43, Cham: Springer.
Barker, Andrew (1989), *Greek Musical Writings, II: Harmonic and Acoustic Theory*, Cambridge: Cambridge University Press.
Baten, Joerg, and Stefan Priwitzer (2015), "Social and Intertemporal Differences of Basic Numeracy in Pannonia (First Century BCE to Third Century CE)," *Scandinavian Economic History Review*, 63 (2): 110–34.
Bausi, Alessandro, Christian Brockmann, Michael Friedrich, and Sabine Kienitz, eds. (2018), *Manuscripts and Archives: Comparative Views on Record-Keeping*, Studies in Manuscript Cultures, #11, Berlin: De Gruyter.
Beaulieu, Paul-Allain (2006), "De l'Esagil au Mouseion: L'organisation de la recherche scientifique au IVe siècle avant J.-C.," *Persika*, 9: 17–36.
Beaumont, Peter B., and Robert T. Bednarik (2013), "Tracing the Emergence of Palaeoart in Sub-Saharan Africa," *Rock Art Research*, 30 (1): 33–54.
Becker, Andrea (2007), "The Royal Game of Ur," in Irving L. Finkel ed., *Ancient Board Games in Perspective: Papers from the 1990 British Museum Colloquium, with Additional Contributions*, 11–15, London: British Museum Press.
Beckmann, Petr (1971), *A History of Pi*, New York: St. Martin's Press.
Bell, Robert C., and Charlotte M. Roueché (2007), "Graeco-Roman Pavement Signs and Game Boards: A British Museum Working Typology," in Irving L. Finkel ed., *Ancient Board Games in Perspective: Papers from the 1990 British Museum Colloquium, with Additional Contributions*, 106–9, London: British Museum Press.
Benedict, Ruth ([1934] 1959), *Patterns of Culture*, Boston, MA: Houghton Mifflin.
Berggren, J. Lennart (1984), "History of Greek Mathematics: A Survey of Recent Research," *Historia Mathematica*, 11 (4): 394–410.
Berggren, J. Lennart (1991), "Ptolemy's Maps of Earth and the Heavens," *Archive for History of Exact Sciences*, 43 (2): 133–44.
Berggren, J. Lennart, and Alexander Jones, trans. and comm. (2000), *Ptolemy's* Geography*: An Annotated Translation of the Theoretical Chapters*, Princeton, NJ: Princeton University Press.
Berggren, J. Lennart, and R. S. D. Thomas (2006), *Euclid's* Phaenomena*: A Translation and Study of a Hellenistic Treatise in Spherical Astronomy*, History of Mathematics, #29, New York/London: American Mathematical Society/London Mathematical Society.
Bernard, Alain, and Christine Proust, eds. (2014), *Scientific Sources and Teaching Contexts throughout History: Problems and Perspectives*, New York: Springer.

Bernard, Alain, Christine Proust, and Micah Ross (2014), "History of Mathematics Education in Antiquity," in Alexander Karp and Gert Schubring eds., *Handbook on the History of Mathematics Education*, 27–53, New York: Springer.
Blåsjö, Viktor (2016), "In Defence of Geometric Algebra," *Archive for History of Exact Sciences*, 70 (3): 325–59.
Bowman, Alan K. (1994), *Life and Letters on the Roman Frontier: Vindolanda and Its People*, London: British Museum Press.
Boyer, Carl B. ([1949] 1959), *The History of the Calculus and its Conceptual Development*, New York: Dover.
Brack-Bernsen, Lis, and John M. Steele (2004), "Babylonian Mathemagics: Two Mathematical Astronomical-Astrological Texts," in Charles Burnett, Jan P. Hogendijk, Kim Plofker, and Michio Yano eds., *Studies in the History of the Exact Sciences in Honour of David Pingree*, 95–135, Leiden: Brill.
Bretelle-Establet, Florence, ed. (2010), *Looking at it from Asia: The Processes that Shaped the Sources of History of Science*, New York: Springer.
Britton, John P., Christine Proust, and Steve Shnider (2011), "Plimpton 322: A Review and a Different Perspective," *Archive for History of Exact Sciences*, 65 (5): 519–66.
Brown, David, John Fermor, and Christopher Walker (1999–2000), "The Water Clock in Mesopotamia," *Archiv für Orientforschung*, 47: 130–48.
Buccellati, Giorgio (2008), "A Horse Training Manual," in Joan Aruz, Kim Benzel, and Jean M. Evans eds., *Beyond Babylon: Art, Trade, and Diplomacy in the Second Millennium B.C.*, 158, New Haven, CT: Yale University Press.
Budge, E. A. Wallis ([1926] 1977), *The Dwellers on the Nile: The Life, History, Religion and Literature of the Ancient Egyptians*, New York: Dover.
Burkert, Walter (1972), *Lore and Science in Ancient Pythagoreanism*, trans. Edwin L. Minar, Jr., Cambridge, MA: Harvard University Press.
Burnett, Charles, Jan P. Hogendijk, Kim Plofker, and Michio Yano, eds. (2004), *Studies in the History of the Exact Sciences in Honour of David Pingree*, Leiden: Brill.
Campbell-Kelly, Martin, Mary Croarken, Raymond Flood, and Eleanor Robson, eds. (2003), *The History of Mathematical Tables: From Sumer to Spreadsheets*, Oxford: Oxford University Press.
Caveing, Maurice (1994), *Essai sur le savoir mathématique dans la Mésopotamie et l'Egypte anciennes*, Lille: Presses universitaires de Lille.
Chakrabarti, Dilip K., and Makkhan Lal, eds. (2014), *History of India, vol. II: Protohistoric Foundations*, New Delhi: Aryan Books International.
Charpin, Dominique (2017), *La vie méconnue des temples mésopotamiens*, Paris: Les Belles Lettres.
Civil, Miguel (1985), "Sur les livres d'écoliers à l'époque paléo-babylonienne," in Jean-Marie Durand and Jean-Robert Kupper eds., *Miscellanea Babylonica, Mélanges offerts à Maurice Birot*, 67–78, Paris: Éditions Recherche sur les Civilisations.
Civil, Miguel, Margaret W. Green, and Wilfred G. Lambert (1979), *Ea A = nâqu, Aa A = nâqu, with their Forerunners and Related Texts*, Materials for the Sumerian Lexicon, #14, Rome: Pontificium Institutum Biblicum.
Clagett, Marshall (1999), *Ancient Egyptian Science: A Source Book, vol. 3: Ancient Egyptian Mathematics*, Philadelphia, PA: American Philosophical Society.
Clancier, Philippe (2014), "Teaching and Learning Medicine and Exorcism at Uruk during the Hellenistic Period," in Alain Bernard and Christine Proust eds., *Scientific Sources and Teaching Contexts throughout History: Problems and Perspectives*, 41–66, New York: Springer.

Clarysse, Willy, and Dorothy J. Thompson (1995), "The Salt-Tax Rate Once Again," *Chronique d'Egypte*, 70 (139/40): 223–9.
Cohen, Morris R., and Israel E. Drabkin (1948), *A Source Book in Greek Science*, Cambridge, MA: Harvard University Press.
Collins, Harry M. (2010), *Tacit and Explicit Knowledge*, London: University of Chicago Press.
Cornell, Tim (1995), *The Beginnings of Rome: Italy and Rome from the Bronze Age to the Punic Wars, c. 1000–264 BC*, London: Routledge.
Corry, Leo (2013), "Geometry and Arithmetic in the Medieval Traditions of Euclid's Elements: A View from Book II," *Archive for History of Exact Sciences*, 67 (6): 637–705.
Couchoud, Sylvia (1993), *Mathématiques égyptiennes: Recherches sur les connaissances mathématiques de l'Egypte pharaonique*, Paris: Le Léopard d'Or.
Crawford, Michael H. (1974), *Roman Republican Coinage*, Cambridge: Cambridge University Press.
Cullen, Christopher (1996), *Astronomy and Mathematics in Ancient China: The Zhou Bi Suan Jing*, Cambridge: Cambridge University Press.
Cuomo, Serafina (2001), *Ancient Mathematics*, London: Routledge.
Cuomo, Serafina (2012), "Exploring Ancient Greek and Roman Numeracy," *Journal of the British Society for the History of Mathematics*, 27 (1): 1–12.
Cuomo, Serafina (2013), "Accounts, Numeracy and Democracy in Classical Athens," in Markus Asper ed., *Writing Science: Medical and Mathematical Authorship in Ancient Greece*, 255–78, Berlin: De Gruyter.
D'Ambra, Eve (1995), "Mourning and the Making of Ancestors in the Testamentum Relief," *American Journal of Archaeology*, 99 (4): 667–81.
Danino, Michel (2014), "Metrology and Linear Measurements in the Harappan Civilization," in Dilip K. Chakrabarti and Makkhan Lal eds., *History of India, vol. II: Protohistoric Foundations*, 309–20, New Delhi: Aryan Books International.
Dauben, Joseph, Shuchun Guo, and Yibao Xu (2013), *Nine Chapters on the Art of Mathematics: A Critical Edition and English Translation Based upon a New Collation of the Ancient Text and Modern Chinese*, 3 vols, Shenyang: Liaoning Education Press.
de Falco, V., M. Krause, and O. Neugebauer (1966), *Hypsikles. Die Aufgangszeiten der Gestirne*, trans. and ed. V. de Falco and M. Kraus, with an introduction by O. Neugebauer, Göttingen: Vandenhoeck & Ruprecht.
Dehaene, Stanislas (2011), *The Number Sense: How the Mind Creates Mathematics*, 2nd edn, New York: Oxford University Press.
Depauw, Mark (1997), *A Companion to Demotic Studies*, Papyrologica Bruxellensia, #28, Bruxelles: Fondation égyptologique rein Elisabeth.
Diels, Hermann, and W. Kranz, eds. (1951–2), *Die Fragmente der Vorsokratiker*, 3 vols, 6th edn, Berlin: Wiedemann.
Dilke, O. A. W. (1987a), "Cartography in the Ancient World: An Introduction," in J. B. Harley and David Woodward eds., *The History of Cartography, vol. 1: Cartography in Prehistoric, Ancient, and Medieval Europe and the Mediterranean*, 105–6 and 276–9, Chicago, IL: University of Chicago Press.
Dilke, O. A. W. (1987b), *Mathematics and Measurement*, Berkeley, CA and London: University of California Press and British Museum.
Donbaz, Veysel (1985), "More Old Assyrian Tablets from Aššur," *Akkadica*, 42: 1–23.
Duncan-Jones, Richard (1977), "Age-Rounding, Illiteracy and Social Differentiation in the Roman Empire," *Chiron*, 7: 333–53.

Durand, Jean-Marie, and Jean-Robert Kupper, eds. (1985), *Miscellanea Babylonica, Mélanges offerts à Maurice Birot*, Paris: Éditions Recherche sur les Civilisations.

Edzard, Dietz O. (1968), "Eine altsumerische Rechentafel (OIP 14, 70)," in W. Röllig and M. Dietrich eds., *Lišān mithurti. Festschrift Wolfram Freiherr von Soden zum 19.VI.1968 gewidemet von Schülern und Mitarbeitern*, Alter Orient und Altes Testament 1, Neukirchen-Vluyn: Neukirchener Verlag des Erziehungsvereins.

Edzard, Dietz O., ed. (1987–90), *Reallexikon der Assyriologie*, Berlin: De Gruyter.

Escobar, Eduardo A. (2018), "'You,' 'Us,' and 'Them': Glass and Procedural Knowledge in Cuneiform Cultures," *History of Knowledge*, December 5. Available online: https://historyofknowledge.net/2018/12/05/you-us-and-them-glass-and-procedural-knowledge-in-cuneiform-cultures/ (accessed February 26, 2023).

Escobar, Eduardo A. (2019), "Glassmaking as Scribal Craft," in Annastasia Amrhein, Clare Fitzgerald, and Elizabeth Knott eds., *A Wonder to Behold: Craftsmanship and the Creation of Babylon's Ishtar Gate*, 119–25, Princeton, NJ: Princeton University Press.

Evans, James (1998), *The History and Practice of Ancient Astronomy*, New York: Oxford University Press.

Evans, James, and J. Lennart Berggren, trans. and comm. (2006), *Geminos's Introduction to the Phenomena*, Princeton, NJ: Princeton University Press.

Everett, Caleb (2017), *Numbers and the Making of Us*, Cambridge, MA: Harvard University Press.

Eves, Howard (1983), *An Introduction to the History of Mathematics*, 5th edn, Philadelphia, PA: Saunders College Publishing.

Faivre, Xavier, Brigitte Lion, and Cécile Michel, eds. (2009), *Et il y eut un esprit dans l'Homme: Jean Bottéro et la Mésopotamie*, Paris: De Boccard.

Fauvel, John, and Jeremy Gray, eds. (1987), *The History of Mathematics: A Reader*, New York: Palgrave Macmillan.

Feliu, Lluís, Fumi Karahashi, and Gonzalo Rubio, eds. (2017), *The First Ninety Years: A Sumerian Celebration in Honor of Miguel Civil*, Boston, MA: De Gruyter.

Finkel, Irving (2000), "On Late Babylonian Medical Training," in Andrew R. George and Irving L. Finkel eds., *Widsom, Gods and Literature: Studies in Honour of W.G. Lambert*, 137–223, Winona Lake, IN: Eisenbrauns.

Finkel, Irving L. (2007a), "On the Rules of the Royal Game of Ur," in Irving. L. Finkel ed., *Ancient Board Games in Perspective: Papers from the 1990 British Museum Colloquium, with Additional Contributions*, 16–32, London: British Museum Press.

Finkel, Irving L., ed. (2007b), *Ancient Board Games in Perspective: Papers from the 1990 British Museum Colloquium, with Additional Contributions*, London: British Museum Press.

Foster, Benjamin R. (2016), *The Age of Agade: Inventing Empire in Ancient Mesopotamia*, New York: Routledge.

Foster, Benjamin R. (2018), *Sargonic Texts from Telloh in the Istanbul Archaeological Museums*, part 2, Atlanta, GA: Lockwood Press.

Frahm, Eckart (2009), *Historische und historisch-literarische Texte aus Assur. KAL 3*, Wiesbaden: Harrassowitz.

Friberg, Jöran (1987–90), "Mathematik," *Reallexicon der Assyriologie*, 7: 531–85.

Friberg, Jöran (2007), *A Remarkable Collection of Babylonian Mathematical Texts*, New York: Springer.

Friberg, Jöran, and Farouk N. H. Al-Rawi (2016), *New Mathematical Cuneiform Texts*, Sources and Studies in the History of Mathematics and Physical Sciences, Cham: Springer.

Fried, Michael N. (2018), "Ways of Relating to Mathematics of the Past," *Journal of Humanistic Mathematics*, 8 (1): 3–23.

Fried, Michael N., and Sabetai Unguru (2001), *Apollonius of Perga's* Conica*: Text, Context, Subtext*, Leiden: Brill.

George, Andrew R. (1992), *Babylonian Topographical Texts*, Orientalia Lovaniensia Analecta, #40, Leuven: Peeters.

George, Andrew R. (2005–6), "The Tower of Babel: Archaeology, History, and Cuneiform Texts," *Archiv für Orientforschung*, 51: 75–95.

George, Andrew R., and Irving L. Finkel, eds. (2016), *Widsom, Gods and Literature: Studies in Honour of W.G. Lambert*, Winona Lake, IN: Eisenbrauns.

Goodman, Nelson (1968), *Languages of Art: An Approach to a Theory of Symbols*, Indianapolis, IN: Bobbs-Merrill.

Gowers, Timothy, ed. (2008), *The Princeton Companion to Mathematics*, Princeton, NJ: Princeton University Press.

Grabiner, Judith V. (2009), "Why did Lagrange 'Prove' the Parallel Postulate?," *American Mathematical Monthly*, 116 (1): 3–18.

Granger, Frank, trans. (1983), Vitruvius: *De Architectura*, vol. 2, Cambridge, MA: Harvard University Press.

Grattan-Guinness, Ivor (2004), "The Mathematics of the Past: Distinguishing its History from Our Heritage," *Historia Mathematica*, 31 (2): 163–85.

Gruber, Martin, and Michael Roaf (2016), "Alternative Interpretations of the Early Mesopotamian Building Plan on RTC 145," *Revue d'assyriologie et d'archéologie orientale*, 110: 35–52.

Guicciardini, Niccolò, ed. (2021), *Anachronisms in the History of Mathematics: Essays on the Historical Interpretation of Mathematical Texts*, Cambridge: Cambridge University Press.

Guy, Richard K. (1988), "The Strong Law of Small Numbers," *American Mathematical Monthly*, 95 (8): 697–712.

Hagen, Frederik (with a contribution by Daniel Soliman) (2018), "Archives in Ancient Egypt, 2500–1000 BCE," in Alessandro Bausi, Christian Brockmann, Michael Friedrich, and Sabine Kienitz eds., *Manuscripts and Archives: Comparative Views on Record-Keeping*, Studies in Manuscript Cultures, #11, 71–170, Berlin: De Gruyter.

Hannah, Robert (2009), *Time in Antiquity*, London: Routledge.

Harley, J. B., and David Woodward, eds. (1987), *The History of Cartography, vol. 1: Cartography in Prehistoric, Ancient, and Medieval Europe and the Mediterranean*, Chicago, IL: University of Chicago Press.

Hartwig, Melinda, ed. (2013), *The Tomb Chapel of Menna: The Art, Culture, and Science of Painting in an Egyptian Tomb*, New York: American University in Cairo Press.

Hasse, Helmut, and Heinrich Scholz (1928), *Die Grundlagenkrisis der griechischen Mathematik*, Berlin: De Gruyter.

Heath, Thomas L., ed. ([1897] 2013), *The Works of Archimedes*, New York: Dover.

Heath, Thomas L. (1913), *Aristarchus of Samo*. Oxford: Clarendon Press.

Heath, Thomas L. (1921), *A History of Greek Mathematics*, 2 vols, Oxford: Clarendon Press.

Heath, Thomas L. (1956), *The Thirteen Books of Euclid's Elements, Translated from the Text of Heiberg with Introduction and Commentary*, 3 vols. New York: Dover.

Heeßel, Nils P. (2018), "A New Medical Therapeutic Text on Rectal Disease," in Strahil V. Panayotov, Luděk Vacín, and Gene Trabich eds., *Mesopotamian Medicine and Magic: Studies in Honor of Markham J. Geller*, 310–42, Leiden: Brill.

Ho, Peng-Yoke (1991), “Chinese Science: The Traditional Chinese View,” *Bulletin of the School of Oriental and African Studies*, 54 (3): 506–19.

Horowitz, Wayne (2014), *The Three Stars Each: The Astrolabes and Related Texts*, Wien: Institut für Orientalistik der Universität Wien.

Høyrup, Jens (1994), *In Measure, Number and Weight*, New York: State University of New York Press.

Høyrup, Jens (2002), *Lengths, Widths, Surfaces: A Portrait of Old Babylonian Algebra and its Kin*, New York: Springer.

Høyrup, Jens (2018), “Mesopotamian Mathematics,” in Alexander Jones and Liba Taub eds., *Cambridge History of Science, vol. 1: Ancient Science*, 59–72, Cambridge: Cambridge University Press.

Huffman, Carl A. (1993), *Philolaus of Croton: Pythagorean and Presocratic*, Cambridge: Cambridge University Press.

Huffman, Carl A. (2005), *Archytas of Tarentum: Pythagorean, Philosopher and Mathematician King*, Cambridge: Cambridge University Press.

Huffman, Carl A., ed. (2014), *A History of Pythagoreanism*, Cambridge: Cambridge University Press.

Hunger, Hermann, and David Pingree (1989), *MUL.APIN: An Astronomical Compendium in Cuneiform*, Archiv für Orientforschung Beiheft, #24, Horn: Verlag Ferdinand Berger & Söhne.

Hunger, Hermann, and John Steele (2019), *The Babylonian Astronomical Compendium MUL.APIN*, Abingdon: Routledge.

Imhausen, Annette (2003), “Egyptian Mathematical Texts and their Contexts,” *Science in Context*, 16 (3): 367–89.

Imhausen, Annette (2010), “Traditions and Myths in the Historiography of Egyptian Mathematics,” in Eleanor Robson and Jacqueline Stedall eds., *The Oxford Handbook of the History of Mathematics*, 781–800, Oxford: Oxford University Press.

Imhausen, Annette (2016), *Mathematics in Ancient Egypt: A Contextual History*, Princeton, NJ: Princeton University Press.

Jacob, Christian, ed. (2007), *Lieux de savoir: Espaces et communautés*, Paris: Albin Michel.

Jakob, Stefan (2003), *Mittelassyrische Verwaltung und Sozialstruktur*, Cuneiform Monographs, #29, Leiden: Brill-Styx.

Jakob, Stefan (2009), *Die mittelassyrischen Texte aus Tell Chuēra in Nordost-Syrien*, Wiesbaden: Harrassowitz.

Jones, Alexander (2005), “Ptolemy’s Canobic Inscription and Heliodorus’ Observation Reports,” *SCIAMVS: Sources and Commentaries in Exact Sciences*, 6: 53–97.

Jones, Alexander (2013), “Canobic Inscription,” in Roger S. Bagnall, Kai Brodersen, Craige B. Champion, Andrew Erskine, and Sabine R. Huebner eds., *The Encyclopedia of Ancient History*, 1299, Malden, MA: Wiley-Blackwell.

Jones, Alexander, and Liba Taub, eds. (2018), *Cambridge History of Science, vol. 1: Ancient Science*, Cambridge: Cambridge University Press.

Kammenhuber, Annelies (1961), *Hippologia Hethitica*, Wiesbaden: Harrassowitz.

Kammenhuber, Annelies (1988), “On Hittites, Mittani-Hurrians, Indo-Aryans and Horse Tablets in the IInd Millennium B.C.,” in Takahito Mikasa ed., *Essays on Anatolian Studies in the Second Millennium B.C.*, 35–51, Wiesbaden: Harrassowitz.

Karp, Alexander, and Gert Schubring, eds. (2014), *Handbook on the History of Mathematics Education*, New York: Springer.

Katz, Victor J., ed. (2007), *The Mathematics of Egypt, Mesopotamia, China, India, and Islam*, Princeton, NJ: Princeton University Press.

Kaye, Noah (2018), "Taxation in the Greco-Roman World: The Hellenistic East," *Oxford Handbooks Online*, Oxford: Oxford University Press. Available online: https://doi.org/10.1093/oxfordhb/9780199935390.013.36.

Kehoe, Dennis P. (1992), *Management and Investment on Estates in Roman Egypt during the Early Empire*, Bonn: Habelt.

Ker, James (2010), "'Nundinae': The Culture of the Roman Week," *Phoenix*, 64 (3/4): 360–85.

Knorr, Wilbur Richard (1975), *The Evolution of the Euclidean Elements*, Dordrecht: D. Reidel.

Knorr, Wilbur Richard (1986), *The Ancient Tradition of Geometric Problems*, Boston, MA: Birkhäuser.

Krause, M., trans. and comm. (1936), *Die Sphärik von Menelaus von Alexandrien*, Abhandlungen der Gesellschaft der Wissenschaften zu Göttingen, #17, Berlin: Weidmannsche buchhandlung.

Kroeber, A. L., and Clyde Kluckhohn ([1952] 1963), *Culture: A Critical Review of Concepts and Definitions*, New York: Alfred A. Knopf.

Kuper, Adam (1999), *Culture: The Anthropologists' Account*, Cambridge, MA: Harvard University Press.

Kurke, Leslie (1999a), "Ancient Greek Board Games and How to Play Them," *Classical Philology*, 94 (3): 247–67.

Kurke, Leslie (1999b), *Coins, Bodies, Games, and Gold: The Politics of Meaning in Archaic Greece*, Princeton, NJ: Princeton University Press.

Lambert, W. G. (2013), *Babylonian Creation Myths*, Winona Lake, IN: Eisenbrauns.

Lang, Mabel (1957), "Herodotos and the Abacus," *Hesperia: Journal of the American School of Classical Studies at Athens*, 26 (3): 271–88.

Lang, Philippa, ed. (2004), *Reinventions: Essays on Hellenistic and Early Roman Science*, Kelowna, BC: Academic Printing and Publishing.

Larsen, Mogens Trolle (2015), *Ancient Kanesh: A Merchant Colony in Bronze Age Anatolia*, New York: Cambridge University Press.

Lear, Jonathan (1982), "Aristotle's Philosophy of Mathematics," *Philosophical Review*, 91 (2): 161–92.

Lecompte, Camille (2020), "The Measurement of Fields During the Pre-Sargonic Period," in Cécile Michel and Karine Chemla eds., *Mathematics, Administrative and Economic Activities in the Ancient Worlds*, Why Sciences of the Ancient World Matter, #5, 283–344, Cham: Springer.

Lehoux, Daryn (2005), "The Parapegma Fragments from Miletus," *Zeitschrift für Papyrologie und Epigraphik*, 152: 125–40.

Lehoux, Daryn (2007), *Astronomy, Weather, and Calendars in the Ancient World: Parapegmata and Related Texts in Classical and Near Eastern Societies*, Cambridge: Cambridge University Press.

Leichty, Erle (2011), *The Royal Inscription of Esarhaddon, King of Assyria (680–669 BC)*, Royal Inscriptions of the Neo-Assyrian Period, #4, Winona Lake, IN: Eisenbrauns.

Lightbody, David I., and Franck Monnier (2017), "An Elegant Vault Design Principle Identified in Old and New Kingdom Architecture," *Journal of Ancient Egyptian Architecture*, 2: 55–69.

Lion, Brigitte (2011), "Literacy and Gender," in Karen Radner and Eleanor Robson eds., *The Oxford Handbook of Cuneiform Culture*, 90–112, Oxford: Oxford University Press.
Livingstone, Alisdair (1986), *Mystical and Mythological Explanatory Works of Assyrian and Babylonian Scholars*, Oxford: Clarendon Press.
Livingstone, Alisdair (1989), *Court Poetry and Literary Miscellanea*, State Archives of Assyria, #3, Helsinki: Helsinki University Press.
Lloyd, G. E. R. (1987), *The Revolutions of Wisdom*, Berkeley, CA: University of California Press.
Lloyd, G. E. R. (1996), *Aristotelian Explorations*, Cambridge: Cambridge University Press.
Lloyd, G. E. R. (2006a), "Mathematics as a Model of Method in Galen," in G. E. R. Lloyd, *Principles and Practices in Ancient Greek and Chinese Science*, Aldershot: Ashgate.
Lloyd, G. E. R. (2006b), *Principles and Practices in Ancient Greek and Chinese Science*, Aldershot: Ashgate.
Lloyd, G. E. R. (2006c), "Theories and Practices of Demonstration in Galen," in G. E. R. Lloyd, *Principles and Practices in Ancient Greek and Chinese Science*, Aldershot: Ashgate.
Maher, David W., and John F. Makowski (2001), "Literary Evidence for Roman Arithmetic with Fractions," *Classical Philology*, 96 (4): 376–99.
Marrou, H. I. (1982), *A History of Education in Antiquity*, trans. George Lamb, Madison, WI: University of Wisconsin Press.
Marshack, Alexander (1991), *The Roots of Civilization*, Mount Kisco, NY: Mayer Bell.
Marzahn, Joachim (2019), "Equids in Mesopotamia: A Short Ride through Selected Textual Sources," in Peter Raulwing, Katheryn M. Linduff, and Joost H. Crouwel eds., *Equids and Wheeled Vehicles in the Ancient World: Essays in Memory of Mary A. Littauer*, 71–84, Oxford: BAR Publishing.
Melchert, Craig H. (1980), "The Use of IKU in Hittite Texts," *Journal of Cuneiform Studies*, 32 (1): 50–6.
Meninger, Karl (1969), *Number Words and Number Symbols: A Cultural History of Numbers*, Cambridge, MA: MIT Press (translation of *Zahlwort und Ziffer*, Göttingen: Vandenhoeck & Ruprecht, 1958).
Michalowski, Piotr (2012), "Literacy Schooling and the Transmission of Knowledge in Early Mesopotamian Culture," in W. S. van Egmond and Wilfred H. van Soldt eds., *Theory and Practice of Knowledge Transfer: Studies in School Education in the Ancient Near East and Beyond. Papers Read at a Symposium in Leiden, 17–19 December 2008*, 39–57, Leiden: Nederlands Instituut voor het Nabije Oosten.
Michel, Cécile (2008), "Ecrire et compter chez les marchands assyriens du début du IIe millénaire av. J.-C.," in Taner Tarhan, Aksel Tibet, and Erkan Konyar eds., *Muhibbe Darga Armağanı*, 345–64, Istanbul: Sadberk Hanım Müzesi.
Michel, Cécile (2020a), "Computation Practices of the Assyrian Merchants during the Nineteenth Century BCE," in C. Michel and K. Chemla eds., *Mathematics, Administrative and Economic Activities in the Ancient Worlds*, Why Sciences of the Ancient World Matter, #5, 399–434, Cham: Springer.
Michel, Cécile (2020b), *Women of Assur and Kanesh: Texts from the Archives of Assyrian Merchants*, Writings from the Ancient World, #42, Baltimore, MD: SBL Press.

Michel, Cécile, and Karine Chemla (2020a), "Mathematics, Administrative and Economic Activities in the Ancient Worlds: An Introduction," in Cécile Michel and Karine Chemla eds., *Mathematics, Administrative and Economic Activities in the Ancient Worlds*, Why Sciences of the Ancient World Matter, #5, 1–51, Cham: Springer.

Michel, Cécile, and Karine Chemla, eds. (2020b), *Mathematics, Administrative and Economic Activities in the Ancient Worlds*, Why Sciences of the Ancient World Matter, #5, Cham: Springer.

Michel, Marianne (2014), *Les mathématiques de l'Egypte ancienne*, Bruxelles: Safran.

Mikasa, Takahito, ed. (1988), *Essays on Anatolian Studies in the Second Millennium B.C.*, Wiesbaden: Harrassowitz.

Miller, Kassandra J., and Sarah L. Symons, eds. (2019), *Down to the Hour: Short Time in the Ancient Mediterranean and Near East*, Leiden: Brill.

Monnier, Franck (2019), "Les pyramides dans les problèmes mathématiques égyptiens," *ENiM*, 12: 233–46.

Morrow, Glenn R. ([1970] 1992), *Proclus: A Commentary on the First Book of Euclid's Elements*, Princeton, NJ: Princeton University Press.

Nemet-Nejat, Karen R. (1993), *Cuneiform Mathematical Texts as a Reflection of Everyday Life in Mesopotamia*, American Oriental Series, #75, New Haven, CT: American Oriental Society.

Netz, Reviel (1999), *The Shaping of Deduction in Greek Mathematics*, Cambridge: Cambridge University Press.

Netz, Reviel (2002), "Counter Culture: Towards a History of Greek Numeracy," *History of Science*, 40 (3): 321–52.

Netz, Reviel (2009), *Ludic Proof*, Cambridge: Cambridge University Press.

Netz, Reviel (2017), *The Works of Archimedes, vol. 3: On Spirals*, Cambridge: Cambridge University Press.

Neugebauer, Otto (1935–7), *Mathematische Keilschrift-Texte*, I–III, Berlin: Springer.

Neugebauer, Otto (1953), "On the 'Hippopede' of Eudoxus," *Scripta Mathematica*, 19 (4): 225–9.

Neugebauer, Otto (1975), *A History of Ancient Mathematical Astronomy*, Berlin: Springer.

Neugebauer, Otto, and Abraham J. Sachs (1945), *Mathematical Cuneiform Texts*, New Haven, CT: American Oriental Series & American Schools of Oriental Research.

Nissen, Hans J., Peter Damerov, and Robert K. Englund (1993), *Archaic Bookkeeping: Writing and Techniques of Economic Administration in the Ancient Near East*, trans. Paul Larsen, Chicago, IL: University of Chicago Press.

Okamoto, Yukari, Mary E. Brenner, and Reagan Curtis (1999), "Numeracy and Culture," in Robert A. Wilson and Frank C. Keil eds., *The MIT Encyclopedia of the Cognitive Sciences*, 611–12, Cambridge, MA: MIT Press.

Oppenheim, A. Leo, Robert H. Brill, Dan Barag, and Axel von Saldern (1970), *Glass and Glassmaking in Ancient Mesopotamia: An Edition of the Cuneiform Texts which Contain Instructions for Glassmakers with a Catalogue of Surviving Objects*, London: Corning Museum of Glass.

Ossendrijver, Mathieu (2012), *Babylonian Mathematical Astronomy: Procedure Texts*, Berlin: Springer.

Ossendrijver, Mathieu (2014), "The Powers of 9 and Related Mathematical Tables from Babylon," *Journal of Cuneiform Studies*, 66: 149–65, with plates.

Ossendrijver, Mathieu (2016), "Translating Babylonian Astronomical Diaries and Procedural Texts," in Annette Imhausen and Tanja Pommerening eds., *Translating Writings of Early Scholars in the Ancient Near East, Egypt, Greece and Rome: Methodological Aspects with Examples*, 125–72. Berlin: De Gruyter.

Ossendrijver, Mathieu (2019), "Scholarly Mathematics in the Rēš Temple," in Christine Proust and John Steele eds., *Scholars and Scholarship in Late Babylonian Uruk*, Why the Sciences of the Ancient World Matter, #2, 187–217, Cham: Springer Nature.

Ostapenko, S. N. (2016), "Bone tessera from Phanagoria," *Problemy istorii, filologii, kul'tury* 4, 71-78. (In Russian).

Panayotov, Strahil V., Luděk Vacín, and Gene Trabich, eds. (2018), *Mesopotamian Medicine and Magic: Studies in Honor of Markham J. Geller*, Leiden: Brill.

Papadopoulos, John K., and Gary Urton, eds. (2012), *The Construction of Value in the Ancient World*, Los Angeles, CA: University of California Press.

Parker, Richard A. (1972), *Demotic Mathematical Papyri*, Providence, RI: Brown University Press/Lund Humphries.

Pausanias (1918), *Description of Greece*, trans. W. H. S. Jones and H. A. Ormerod, 4 vols, Cambridge, MA/London: Harvard University Press/Heinemann. Available online: http://data.perseus.org/citations/urn:cts:greekLit:tlg0525.tlg001.perseus-eng1:10.31 (accessed January 10, 2020).

Piccione, Peter A. (2007), "The Egyptian Game of Senet and the Migration of the Soul," in Irving L. Finkel ed., *Ancient Board Games in Perspective: Papers from the 1990 British Museum Colloquium, with Additional Contributions*, 54–63, London: British Museum Press.

Playfair, John (1795), *Elements of Geometry*, London: Printed for Bell & Bradfute, and G. G. & J. Robinson.

Pletser, Vladimir (2012), "Does the Ishango Bone Indicate Knowledge of the Base 12? An Interpretation of a Prehistoric Discovery, the First Mathematical Tool of Humankind." Available online: https://doi.org/10.48550/arXiv.1204.1019 (accessed March 5, 2020).

Pollock, Sheldon (1985), "The Theory of Practice and the Practice of Theory in Indian Intellectual History," *Journal of the American Oriental Society*, 105 (3): 499–519.

Pólya, George (1973), *How to Solve It*, 2nd edn, Princeton, NJ: Princeton University Press.

Powell, Marvin A. (1987–90), "Masse und Gewichte," in Dietz. O. Edzard ed., *Reallexikon der Assyriologie*, 457–517, Berlin: De Gruyter.

Proust, Christine (2007), *Tablettes mathématiques de Nippur*, Varia Anatolica, #18, Istanbul: Institut Français d'Etudes Anatoliennes/De Boccard.

Proust, Christine (2008), *Tablettes mathématiques paléo-babyloniennes de la collection Hilprecht*, Texte und Materialen der Hilprecht Collection, #8, Leipzig: Harrassowitz.

Proust, Christine (2010), "Mesopotamian Metrological Lists and Tables: Forgotten Sources," in Florence Bretelle-Establet ed., *Looking at it from Asia: The Processes that Shaped the Sources of History of Science*, 245–76, New York: Springer.

Proust, Christine (2014), "Does a Master Always Write for His Students Some Evidence from Old Babylonian Scribal Schools," in Alain Bernard and Christine Proust eds., *Scientific Sources and Teaching Contexts throughout History: Problems and Perspectives*, 69–94, New York: Springer.

Proust, Christine (2015), "La chanson des mathématiques dans l'*Edubba* (Mésopotamie, début du deuxième millénaire avant notre ère)," *Archives Internationales d'Histoire des Sciences*, 65 (175): 517–47.

Proust, Christine (2019), "A Mathematical Collection Found in the 'House of the āšipus'. The Art of Metrology in Achaemenid Uruk," in Christine Proust and John Steele eds., *Scholars and Scholarship in Late Babylonian Uruk*, Why the Sciences of the Ancient World Matter, #2, 89–146, Cham: Springer Nature.

Proust, Christine (2020), "Early-Dynastic Tables from Southern Mesopotamia, or the Multiple Facets of the Quantification of Surfaces," in Cécile Michel and Karine Chemla eds., *Mathematics, Administrative and Economic Activities in the Ancient Worlds*, Why Sciences of the Ancient World Matter, #5, 345–96, Cham: Springer.

Proust, Christine, and John Steele (2019a), "Introduction: Scholars, Scholarly Archives and the Practice of Scholarship in Late Babylonian Uruk," in Christine Proust and John Steele eds., *Scholars and Scholarship in Late Babylonian Uruk*, Why the Sciences of the Ancient World Matter, #2, 1–52, Cham: Springer Nature.

Proust, Christine, and John Steele, eds. (2019b), *Scholars and Scholarship in Late Babylonian Uruk*, Why the Sciences of the Ancient World Matter, #2, Cham: Springer Nature.

Purcell, Nichholas (2007), "Inscribed Imperial Roman Gaming-Boards," in Irving L. Finkel ed., *Ancient Board Games in Perspective: Papers from the 1990 British Museum Colloquium, with Additional Contributions*, 90–7, London: British Museum Press.

Radner, Karen, and Eleanor Robson, eds. (2011), *The Oxford Handbook of Cuneiform Culture*, New York: Oxford University Press.

Rathbone, Dominic (1991), *Economic Rationalism and Rural Society in Third-Century A.D. Egypt: The Heroninos Archive and the Appianus Estate*, Cambridge: Cambridge University Press.

Rathbone, Dominic (1997), "Prices and Price Formation in Roman Egypt," in Jean Andreau, Pierre Briant, and Raymond Descat eds., *Economie antique: Prix et formation des prix dans les économies antiques*, 183–244, Toulouse: Saint-Bertrand-de-Comminges.

Raulwing, Peter (2005), "The Kikkuli Text (CTH 284): Some Interdisciplinary Remarks on Hittite Training Texts for Chariot Horses in the Second Half of the 2nd Millennium B.C.," in *Les Équidés dans le monde méditerranéen antique. Actes du colloque organisé par l'École française d'Athènes, le Centre Camille Jullian, et l'UMR 5140 du CNRS, Athènes, 26–28 Novembre 2003*, 61–75, Lattes: Édition de l'Association pour le développement de l'archéologie en Languedoc-Roussillon.

Raulwing, Peter, Katheryn M. Linduff, and Joost H. Crouwel, eds. (2019), *Equids and Wheeled Vehicles in the Ancient World: Essays in Memory of Mary A. Littauer*, Oxford: BAR Publishing.

Remijsen, Sofie (2007), "The Postal Service and the Hour as a Unit of Time in Antiquity," *Historia*, 56 (2): 127–40.

Remmert, Volker R., Martina R. Schneider, and Henrik K. Sørensen, eds. (2016), *Historiography of Mathematics in the Nineteenth and Twentieth Centuries*, Cham: Springer.

Ritter, Jim (2009) "Science and Reason in Ancient Mesopotamia," in Xavier Faivre, Brigitte Lion, and Cécile Michel eds., *Et il y eut un esprit dans l'Homme: Jean Bottéro et la Mésopotamie*, 83–103, Paris: De Boccard.

Robins, Gay, and Charles Shute (1987), *The Rhind Mathematical Papyrus*, London: British Museum Publications.

Robson, Eleanor (1999), *Mesopotamian Mathematics, 2100–1600 BC: Technical Constants in Bureaucracy and Education*, Oxford Editions of Cuneiform Text, #14, Oxford: Clarendon Press.

Robson, Eleanor (2001a), "Technology in Society: Three Textual Case Studies from Late Bronze Age Mesopotamia," in Andrew J. Shortland ed., *The Social Context of Technological Change: Egypt and the Near East, 1650–1550 BC*, 39–57, Oxford: Oxbow Books.

Robson, Eleanor (2001b), "The Tablet House: A Scribal School in Old Babylonian Nippur," *Revue d'Assyriologie*, 95 (1): 39–66.

Robson, Eleanor (2003), "Tables in Sumer, Babyonia, Assyria, 2500 BCE–50 CE," in Martin Campbell-Kelly, Mary Croarken, Raymond Flood, and Eleanor Robson eds., *The History of Mathematical Tables: From Sumer to Spreadsheets*, 19–48, Oxford: Oxford University Press.

Robson, Eleanor (2007a), "Mesopotamian Mathematics," in Victor J. Katz ed., *The Mathematics of Egypt, Mesopotamia, China, India, and Islam*, 58–186, Princeton, NJ: Princeton University Press.

Robson, Eleanor (2007b), "Secrets de famille: Prêtres et astronomes à Uruk à l'époque hellénistique," in Christian Jacob ed., *Lieux de savoir: Espaces et communautés*, 440–60, Paris: Albin Michel.

Robson, Eleanor (2008a), *Mathematics in Ancient Iraq: A Social History*, Princeton, NJ: Princeton University Press.

Robson, Eleanor (2008b), "Numeracy," in Timothy Gowers ed., *The Princeton Companion to Mathematics*, 983–91, Princeton, NJ: Princeton University Press.

Robson, Eleanor (2009), "Mathematics Education in an Old Babylonian Scribal School," in Eleanor Robson and Jacqueline Stedall eds., *The Oxford Handbook of the History of Mathematics*, 199–227, Oxford: Oxford University Press.

Robson, Eleanor, and Jacqueline Stedall, eds. (2009), *The Oxford Handbook of the History of Mathematics*, Oxford: Oxford University Press.

Rochberg, Francesca (2004), *The Heavenly Writing: Divination, Horoscopy and Astronomy in Mesopotamian Culture*, Cambridge: Cambridge University Press.

Rochberg, Francesca (2010), *In the Path of the Moon: Babylonian Celestial Divination and its Legacy*, Leiden: Brill.

Rochberg, Francesca (2018), "Reasoning, Representing, and Modeling in Babylonian Astronomy," *Journal of Near Eastern History*, 5: 131–48.

Rochberg-Halton, Francesca (1984), "New Evidence for the History of Astrology," *Journal of Near Eastern Studies*, 43 (2): 115–40.

Rossi, Corinna (2003), *Architecture and Mathematics in Ancient Egypt*, Cambridge: Cambridge University Press.

Rossi, Corinna (2007), *Architecture and Mathematics in Ancient Egypt*, Cambridge: Cambridge University Press.

Rost, Stephanie (2020), "Insights into the Administration of Ancient Irrigation Systems in Third Millennium BCE Mesopotamia,: in Cécile Michel and Karine Chemla eds., *Mathematics, Administrative and Economic Activities in the Ancient Worlds*, Why Sciences of the Ancient World Matter, #5, 159–200, Cham: Springer.

Sambursky, Samuel (1962), *The Physical World of Late Antiquity*, London: Routledge & Kegan Paul.

Sauvage, Martin (1998), "La construction des ziggurats sous la troisième dynastie d'Ur," *Iraq*, 60: 45–63.

Sauvage, Martin (2020), "Mathematical Computations in the Management of Public Construction Work in Mesopotamia (End of the Third and Beginning of the Second Millennium BCE)," in Cécile Michel and Karine Chemla eds., *Mathematics, Administrative and Economic Activities in the Ancient Worlds*, Why Sciences of the Ancient World Matter, #5, 201–38, Cham Springer Nature.
Schaffer, Leigh S. (2010), "Beyond Berger and Luckmann's Concept of 'Recipe Knowledge': Simple versus Standardized Recipes," *Sociological Viewpoints*, Spring Issue: 55–63.
Schaps, David M. (2004), *The Invention of Coinage and the Monetization of Ancient Greece*, Ann Arbor, MI: Michigan University Press.
Schärlig, Alain (2001), *Compter avec des cailloux: Le calcul élémentaire sur l'abaque chez les anciens Grecs*, Lausanne: Presses Polytechniques et Universitaires Romandes (PPUR).
Schatzberg, Eric (2018), *Technology: Critical History of a Concept*, Chicago, IL: University of Chicago Press.
Scheidel, Walter (1996), *Measuring Sex, Age and Death in the Roman Empire: Explorations in Ancient Demography*, Journal of Roman Archaeology Supplementary Series, #21, Ann Arbor, MI: University of Michigan Press.
Schneider, Martina R. (2016), "Contextualizing Unguru's 1975 Attack on the Historiography of Ancient Greek Mathematics," in Volker R. Remmert, Martina R. Schneider, and Henrik K. Sørensen eds., *Historiography of Mathematics in the Nineteenth and Twentieth Centuries*, 245–67, Cham: Springer.
Schöne, Hermann et al. (1914), *Heronis Alexandrini Opera quae supersunt omnia*, vols I, II.1, III–V, Leipzig: Teubner.
Schuster-Brandis, Anais (2008), *Steine als Schutz- und Heilmittel: Untersuchung zu ihrer Verwendung in der Beschwörungskunst Mesopotamiens im 1. Jt. v. Chr.*, Alter Orient und Altes Testament, #46, Münster: Ugarit-Verlag.
Seaford, Richard (2004), *Money and the Early Greek Mind: Homer, Philosophy, Tragedy*, Cambridge: Cambridge University Press.
Sebbane, Michael (2004), "An Incised Senet Board Game," in David Ussishkin ed., *The Renewed Archaeological Excavations at Lachish*, vol. 2, 690–4, Tel Aviv: Emery and Claire Yass Publications in Archaeology.
Sedley, David N. (2007), *Creationism and its Critics in Antiquity*, Berkeley, CA: University of California Press.
Shortland, Andrew J., ed. (2001), *The Social Context of Technological Change: Egypt and the Near East, 1650–1550 BC*, Oxford: Oxbow Books.
Sidoli, Nathan (2018), "Greek Mathematics," in Alexander Jones and Liba Taub eds., *Cambridge History of Science, vol. 1: Ancient Science*, 345–73, Cambridge: Cambridge University Press.
Sidoli, Nathan, and J. Lennart Berggren (2007), "The Arabic Version of Ptolemy's *Planisphere* or *Flattening the Surface of the Sphere*: Text, Translation Commentary," *SCIAMUS: Sources and Commentaries in Exact Sciences*, 8: 37–140.
Simpson, William K. (2003), *The Literature of Ancient Egypt: An Anthology of Stories, Instructions, Stelae, Autobiographies, and Poetry*, New Haven, CT: Yale University Press.
Sjöberg, Åke W. (1973), "Der Vater und sein missratener Sohn," *Journal of Cuneiform Studies*, 25: 105–50.
Slater, Niall W., ed. (2017), *Voice and Voices in Antiquity: Orality and Literacy in the Ancient World*, Leiden: Brill.

Smith, David Eugene (1925), *History of Mathematic, vol. 2: Special Topics of Elementary Mathematics*, Boston, MA: Ginn.

Sriraman, Bharath, ed. (2019), *Handbook of the Mathematics of the Arts and Sciences*, Cham: Springer.

Steele, John M. (2007), "Celestial Measurement in Babylonian Astronomy," *Annals of Science*, 64 (3): 293–325.

Steele, John M. (2017), *Rising Times Schemes in Babylonian Astronomy*, Springer Briefs in the History of Science and Technology, Berlin: Springer.

Steele, John (2019a), "Astronomical Activity in the 'House of the *āšipu*s' in Uruk," in Christine Proust and John Steele eds., *Scholars and Scholarship in Late Babylonian Uruk*, Why the Sciences of the Ancient World Matter, #2, 14–70, Cham: Springer Nature.

Steele, John (2019b), "Short Time in Mesopotamia," in Kassandra J. Miller and Sarah L. Symons eds., *Down to the Hour: Short Time in the Ancient Mediterranean and Near East*, 90–124, Leiden: Brill.

Strudwick, Nigel (2005), *Texts from the Pyramid Age*, Writings from the Ancient World, #16, Atlanta, GA: Society of Biblical Literature.

Tait, W. John (2007), "Were there Gamesters in Pharaonic Egypt?," in Irving L. Finkel ed., *Ancient Board Games in Perspective: Papers from the 1990 British Museum Colloquium, with Additional Contributions*, 46–53. London: British Museum Press.

Tanret, Michel (2002), *Per aspera ad astra. L'apprentissage du cunéiforme à Sippar-Amnanum pendant la période paléo-babylonienne tardive*, Mesopotamian History and Environment Texts, #1/2, Gent: Universiteit Gent.

Tarhan, Taner, Aksel Tibet, and Erkan Konyar, eds. (2008), *Muhibbe Darga Armağanı*, Istanbul: Sadberk Hanım Müzesi.

Taub, Liba (2017), *Science Writing in Greco-Roman Antiquity*, Cambridge: Cambridge University Press.

Taub, Liba, ed. (2020), *The Cambridge Companion to Ancient Greek and Roman Science*, Cambridge: Cambridge University Press.

Thomas, Ivor (1939–41) *Selections Illustrating the History of Greek Mathematics*, 2 vols, Loeb Classical Library. Cambridge, MA/London: Harvard University Press/ Heinemann.

Thomas, Ivor (1941), *Greek Mathematical Works*, vol. 2, Cambridge, MA: Harvard University Press.

Thureau-Dangin, François (1938), *Textes Mathématiques Babyloniens*, Leiden: Ex Oriente Lux.

Toomer, G. J. (1984), *Ptolemy's* Almagest, New York: Springer.

Tybjerg, K. (2004), "Hero of Alexandria's Mechanical Geometry," in Philippa Lang ed., *Reinventions: Essays on Hellenistic and Early Roman Science*, 29–56, Kelowna, BC: Academic Printing and Publishing.

Tylor, Edward B. ([1871] 1929), *Primitive Culture: Researches into the Development of Mythology, Philosophy, Religion, Language, Art, and Custom*, 2 vols, London: John Murray.

Unguru, Sabetei (1975), "On the Need to Rewrite the History of Greek Mathematics," *Archive for History of Exact Sciences*, 15 (1): 67–114.

Unguru, Sabetei (1979), "History of Ancient Mathematics: Some Reflections of the State of the Art," *Isis*, 70 (4): 555–65.

Ussishkin, David, ed. (2004), *The Renewed Archaeological Excavations at Lachish*, vol. 2, Tel Aviv: Emery and Claire Yass Publications in Archaeology.

van Berkel, Tazuko Angela (2017), "Voiced Mathematics: Orality and Numeracy," in Niall W. Slater ed., *Voice and Voices in Antiquity: Orality and Literacy in the Ancient World*, 321–50, Leiden: Brill.

van Egmond, W. S., and Wilfred H. van Soldt, eds. (2012), *Theory and Practice of Knowledge Transfer: Studies in School Education in the Ancient Near East and Beyond: Papers Read at a Symposium in Leiden, 17–19 December 2008*, Leiden: Nederlands Instituut voor het Nabije Oosten.

Veldhuis, Niek C. (1997), "Elementary Education at Nippur: The Lists of Trees and Wooden Objects," PhD diss., University of Groningen. Available online: https://www.rug.nl/research/portal/publications/pub(d7539f72-9851-4b83-841f-886ef19cb1be).html (accessed September 20, 2019).

von Reden, Sitta (2007), *Money in Ptolemaic Egypt: From the Macedonian Conquest to the End of the Third Century BC*, Cambridge: Cambridge University Press.

Voogt, Alex de, Anne-Elizabeth Dunn-Vaturi, and Jelmer W. Eerkens (2013), "Cultural Transmission in the Ancient Near East: Twenty Squares and Fifty-Eight Holes," *Journal of Archaeological Science*, 40 (4): 1715–30.

Wagner, David L. (1986), *The Seven Liberal Arts in the Middle Ages*, Bloomington, IN: Indiana University Press.

West, Martin, trans. (1988), *Theogony and Works and Day*, New York: Oxford University Press.

Wicksted, P., and Francis Cornford, trans. and eds. (1957), *Aristotle Physics*, books I–IV, Cambridge, MA: Harvard University Press.

Wilder, Raymond L. (1981), *Mathematics as a Cultural System*, Oxford: Pergamon Press.

Williams, Burma P., and Richard S. Williams (1995), "Finger Numbers in the Greco-Roman World and the Early Middle Ages," *Isis*, 86 (4): 587–608.

Wilson, Robert A., and Frank C. Keil, eds. (1999), *The MIT Encyclopedia of the Cognitive Sciences*, Cambridge, MA: MIT Press.

Woods, Christopher (2017), "The Abacus in Mesopotamia: Considerations from a Comparative Perspective.," in Lluís Feliu, Fumi Karahashi, and Gonzalo Rubio eds., *The First Ninety Years: A Sumerian Celebration in Honor of Miguel Civil*, 416–78, Berlin: De Gruyter.

Zaslavsky, Claudia (1992), "Women as the First Mathematicians," *International Study Group on Ethnomathematics Newsletter*, 7 (1). Available online: https://web.nmsu.edu/~pscott/isgem71.htm (accessed March 5, 2020).

Zhmud, Leonid (1998), "Plato as 'Architect of Science,'" *Phronesis*, 43 (3): 211–44.

Zhmud, Leonid (2020), "Ancient Greek Historiography of Science," in Liba Taub ed., *The Cambridge Companion to Ancient Greek and Roman Science*, 268–87, Cambridge: Cambridge University Press.

INDEX

Locators in *italics* refer to figures or illustrations